Joseph Chuen-huei Huang

Participatory Design for Prefab House

Joseph Chuen-huei Huang

Participatory Design for Prefab House

Using Internet and Query Approach of Customizing Prefabricated Houses

VDM Verlag Dr. Müller

Imprint

Bibliographic information by the German National Library: The German National Library lists this publication at the German National Bibliography; detailed bibliographic information is available on the Internet at http://dnb.d-nb.de.

Any brand names and product names mentioned in this book are subject to trademark, brand or patent protection and are trademarks or registered trademarks of their respective holders. The use of brand names, product names, common names, trade names, product descriptions etc. even without a particular marking in this works is in no way to be construed to mean that such names may be regarded as unrestricted in respect of trademark and brand protection legislation and could thus be used by anyone.

Cover image: www.purestockx.com

Publisher:
VDM Verlag Dr. Müller Aktiengesellschaft & Co. KG , Dudweiler Landstr. 125 a, 66123 Saarbrücken, Germany,
Phone +49 681 9100-698, Fax +49 681 9100-988,
Email: info@vdm-verlag.de

Zugl.: Chicago, Illinois Institute of Technology, Diss. , 2008

Produced in USA and UK by:
Lightning Source Inc., La Vergne, Tennessee, USA
Lightning Source UK Ltd., Milton Keynes, UK
BookSurge LLC, 5341 Dorchester Road, Suite 16, North Charleston, SC 29418, USA

ISBN: 978-3-639-07528-1

ACKNOWLEDGEMENT

I would like to express my deep appreciation and thanks to all the individuals who helped and encouraged me to accomplish this work. Especially, I would like to thank my wife, Yio-Hsien, for her love, care, understanding, sacrificing, and continuous encouragement during the entire process. Also, I would like to thank my parents, who believe my strong interest and capabilities in architecture design and research. Moreover, I would like to thank my thesis committee professors for all of the greatest help. Without their knowledge and guidance, I would not have completed this book. I specially thank Professor Mahjoub Elnimeiri, the Director of the PhD in Architecture Program at Illinois Institute of Technology, for encouraging me to see another world of architecture with technology applications. I feel deeply grateful to my adviser, Professor Robert J. Krawczyk, for his knowledgeable advice and energetic motivation in design computing, digital technology, and modular house concept development. Without his inspiration, I could not focus on my research direction in the early phase. I also feel deeply thankful to Professor Robert Babbin, Professor David Sharpe, and Professor George Schipporeit, for their tremendous experience sharing and correction of my work. The last but not least, Professor Keiichi Sato from the Institute of Design, gave me precious help on research methodology and encourage me to make a field trip to see the prefabricated housing park and housing information center in Japan.

Joseph Huang

Chicago, Illinois
August 2008

i

TABLE OF CONTENTS

LIST OF TABLES

LIST OF FIGURES

vi

CHAPTER 1

INTRODUCTION

The design of industrialized housing has been a pre-occupation in architecture since the start of the industrial revolution in the nineteenth century. In the first half of the twentieth century, architects attempted to solve the housing shortage by introducing a production process based on the assembly line. The assembly line was initially developed for the automobile industry by Henry Ford, but soon became a paradigm for the housing industry (Duarte, 2001). For example, the Dymaxion House (Figure 1.1) designed by Buckminster Fuller was trying to achieve the mass production goal by retooling the aircraft factory.

Figure 1.1. Dymaxion House

Prefabrication technology groups building components into larger-scale modular units, such as a prefab wall panel with window and door openings. Each module is made in the factory using assembly line techniques, and then transported to the building site to be installed on a permanent foundation. This is advantageous because it shifts portions of the construction process from the site to the factory where worker productivity is increased, quality is higher, costs are lower, and the overall need for labor is reduced. The construction of a new site-built home in the U.S. typically consists of 80% field labor and 20% material costs[1] – an extraordinarily high labor component compared to other industries. With prefabrication technology, the improvements of quality and efficiency are accomplished because factories can offer better working conditions, automation of many tasks, fewer scheduling and weather-related problems, and simplified inspection processes.

The Sears Roebuck catalogue made prefabricated homes available to subscribers as early as 1908 (Thornton, 2004), and prefabrication was later explored by such eminent twentieth-century architects as Le Corbusier, Walter Gropius, Frank Lloyd Wright, Jean Prouvé, and **Paul Rudolph**, who saw the technology as a new solution to the problem of housing in modern society. After World War II, this approach was extensively used in the reconstruction of Europe and for the postwar housing needs of the United States. Several aircraft companies turned to producing industrialized housing and component parts. Once the housing shortage was satisfied, the implied degree of repetition became unacceptable by a society increasingly focused on individual freedom and choice (Duarte, 2001).

If mass production and prefabrication methods of the assembly line were the ideal of architecture in the early twentieth century, then mass customization and the development of digital technology are the recently emerged paradigms of the twenty-first century. The development of the digital revolution has already prompted the shift towards mass customization. In this new industrial model, the computer-aided manufacturing facilitates variations of the same product. Mass production was all about the economy of making things in quantity, but mass customization does not depend on serial repetitions to be cost effective.

---

[1] Kent Larson, Stephen Intille, Thomas McLeish, Jennifer Beaudin, and Reid Williams, "Open Source Building: Reinventing Places of Living", BT Technology Journal Vol. 22 No. 4, October 2004, pp.198.

It is about cultural production as opposed to the industrial output of mass production[2] (Kieran and Timberlake, 2004). Within limited design parameters, it is possible for customers to determine what options they wish by participating in the flow of the design process from the beginning. This concept has already been implemented in the computer (Dell), clothing (Lands' End), and shoe (Nike) industries, but it has not been fully integrated in the housing industry. The fundamental premise of mass customization is to no longer manufacture products "blindly" according to a predicted demand, but instead allowing production to be directly driven by actual orders (Schodek, 2004).

Today's information technology has become even more interactive and powerful than at the end of the last century. Integrating a participatory home design concept with web technology to create an online interface can become the design platform by which the clients can make more choices and establish a better communication with architects and/or manufacturers. Face-to-face meeting time between architect and client is always limited and time consuming, while a computational web-based design approach is infinitely patient and always available (Larson, Tapia, & Duarte, 2001). One of the problems that prefabricated housing industries failed to address in the twentieth century was the lack of variability and an individual identified design[3] (Kieran and Timberlake, 2004). How prefabricated housing design can be evolved from mass repetitive production level to mass customization level to meet flexibility and variability is the primary issue to be explored in this research.

This book is organized in six chapters. Chapter 1 provides an overview of prefabricated housing and the emerging concept of mass customization by digital technology. The introduction creates a big picture of the selected subject and addresses some current

---

[2] Stephen Kieran and James Timberlake, Refabricating Architecture: How Manufacturing Methodologies are Poised to Transform Building Construction (New York: McGraw-Hill, 2004), pp. 111.
[3] Stephen Kieran and James Timberlake, pp. 108.

3

issues as the motivation of the research. Chapter 2 provides a clear definition of terminologies in this field, and describes history and modular types of prefabricated housing. Besides that, this chapter also reviews prior and current researches, prototypes, and commercial products that support customization of residential design by the end-users with the help of digital applications. After comprehensive literature review, Chapter 3 identifies the missing parts and limitation from the existing research achievements, and sets the research objectives for the further investigation. Chapter 4 discusses the methodology that has to be developed for achieving the goal of customizing prefabricated modular houses by the consumers in the conceptual model and system workflow. Design criteria are also proposed in this chapter as the basis for the prototype design. Further analysis of existing modular house systems and configuration study of selected units are prepared for prototype testing in the next chapter. Chapter 5 presents a series of condition-based case studies, and highlights the development process of digital questionnaire and design configurations. Step by step design interface screenshots demonstrate the input and output simulations of the prototype and integration of the other current digital design applications. Options of professional helps and tools for collaboration are also introduced in this chapter. Finally, Chapter 6 summarizes the research achievements and limitations. The possible directions for future research have been illustrated at the end of this book.

CHAPTER 2

BACKGROUND AND LITERATURE REVIEW

2.1   Definition of Different Terminologies

Prefabricated housing is a general term that indicates modular building components

that are pre-made in the factory, and then transported to the building site to be assembled and

installed on a permanent foundation. It may include manufactured housing (following HUD

code), modular housing (following local zoning and building codes) and production housing

(site-built housing produced in a systematic manner). Each name change reveals a different

categorization system created by the authorities. Table 2.1 includes the definition (based on

North America) and example of each term, and Figure 2.1, a diagram for the different

categories of prefabrication by the type of usage. The background review will focus on the

residential part of prefabrication, especially for analyzing the modularity of prefabricated

housing.

Table 2.1. Definition of Terminologies in Prefabricated Housing (Page 1 of 2)

Term	Definition	Example
Prefabrication	To describe any manufacturing process that takes place in a controlled environment, usually a factory. Its slang version – prefab – is currently in vogue, and while it is applied to many things, it differentiates none of them.	General term
Manufactured Housing	The factory-made home must be legally classified as real property and permanently affixed to a foundation with the characteristics of site-built housing. Since July 15, 1976, all individual sections must display a red shield, which certifies that the manufacturer met all HUD code.	Clayton Manufactured Homes
Mobile Home(s)	Housing made in a factory and transported to a building site which either a permanent or temporary location and hooked up to existing utilities. Sometimes grouped with other mobile homes in mobile home parks.	Double Wides; Trailer Home

5

Table 2.1. Definition of Terminologies in Prefabricated Housing (Page 2 of 2)

Term	Definition	Example
Modular Housing	Dwelling units constructed from components prefabricated in a factory and erected on the site. The structure must assume characteristics of site-built housing and must meet local zoning and building codes.	Habitat '67 by Moshe Safdie, 1967
Panelized House	Panelized or Kit houses were popularized by the Sears, Roebuck Company in 1908, when it began selling do-it-yourself house kits. Panelized factory-built walls are inserted into a modified post-and-beam structure by a builder on-site.	Kit House, Tilt-Up Concrete Panel House
Precut House	House made by precut timber with interlocking wedge-shaped joint.	Log House
Emergency House	Immediate relief in emergencies triggered by natural disaster or war.	Paper Log House by Shigeru Ban, 1995
Container Home	Modified shipping container as modular/ transportable living spaces.	MUD by Lot-ek, 2003

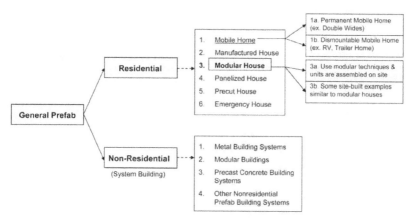

Figure 2.1. Different Categories of Prefabrication by the Type of Usage

2.2 Historical Overview of Prefabricated Housing

Prefabricated housing is not temporary shelter. The majority of prefab housing components are manufactured in the factory, and then these prefab elements are shipped and assembled on the building site with a permanent foundation. Temporary shelter is another

6

category of housing type, and usually relates with the specific ways of living. In the past, the houses of nomadic tribes, like the Native American Tepee and Mongolian Ger, with their flexible construction, allowing them to migrate from one place to another, and to find fresh land for animal feed (Kronenburg, 2002).

The history of prefab housing began nearly four hundred years ago, when a panelized wood house was shipped from England to Cape Ann, Massachusetts in 1624 to provide housing for a fishing fleet (Arieff and Burkhart, 2002). Swedes introduced a notched building-corner technique for the construction of log cabins just a little over a decade later. By the nineteenth century, portable structures had grown in number as new settlements and colonies were formed to support a demand for immediate housing solutions. The kit houses shipped by rail during California Gold Rush in 1849 are one example (Arieff and Burkhart, 2002). During the early part of the twentieth century, many architects and inventors were experimenting with these systems for housing. Table 2.2 summarizes the major developments of prefabricated housing from the beginning of the twentieth century to the present[4].

Table 2.2. Prefabricated Housing Development Timeline from 1906-2007 (Page 1 of 5)

Year	Designer/ Influencer/ Major Achievement
1906	Aladdin Readi - Cut Houses produces a kit house of numbered, precut pieces.
1908	Sears Roebuck & Co. Houses by Mail program established. 100,000 units sold by its demise in 1940.
1923	Walter Gropius and Adolf Meyer develop "Building Blocks," a standardized system of housing.
1929	Buckminster Fuller introduces an early concept for the Dymaxion House—his round metal house—at Chicago's Marshall Fields department store.

---

[4] Most data quote from the website of Some Assembly Required: Contemporary Prefabricated Houses exhibition in Walker Art Center. http://design.walkerart.org/prefab/Main/PrefabTimeline

Table 2.2. Prefabricated Housing Development Timeline from 1906-2007 (Page 2 of 5)

Year	Designer/ Influencer/ Major Achievement
1931	Albert Frey and A. Lawrence Kocher debut the Aluminaire, the first lightweight steel and aluminum house in the U.S.
1932	General Houses Corporation introduces a press-steel panel house for $3,000-$4,500.
	American Houses, Inc. introduces the American Motohome, a simple, box-like, turnkey, steel-framed house.
1935	Wally Byam introduces his iconic, aluminum shell Airstream "Clipper," a trailer easily towed by an automobile.
1936	Frank Lloyd Wright proposes his Usonian House, a system of standardized details and modular dimensions. Although not technically a prefab house, more than 100 are built over the years.
1940	Engineers Peter Dejongh and Otto Brandenberger design the Quonset Hut, a semi-cylindrical structure formed by a ribbed metal shell.
1942	General Panel Corporation commissions Walter Gropius and Konrad Wachsmann to design a panelized house.
1944	Founded in 1944 as a saw mill, the leading Swedish manufacturer - Älvsbyhus started production of prefabricated homes in 1960 and is now the largest manufacturer in the Nordic region.
1945	Developer and builder William Levitt begins Levittown construction. His traditional stick-built, high-volume house assembly method rivals projected prefabricated housing volumes. By 1948 he was finishing 150 houses per week.
	Lindal Cedar Homes established. Using a wooden post and beam system, Lindal offers a customizable and complete kit home package.
1947	Industrial designer Henry Dreyfus and architect Edward Larrabee Barnes collaborate on the design of a prefab house for Vultex Aircraft Company consisting of paper core panels skinned in aluminum.
	John Bemis, a MIT School of Architecture graduate, founds Acorn Structures, a prefabricated building system.
1947	Sekisui Chemical Company was established as Sekisui Sangyo in Osaka, Japan in 1947. After World War II, the reconstruction efforts in the economic, industrial, and political sectors provided houses and job opportunities for millions homeless and soldiers. Sekisui Chemical Company operates three main business divisions, including Housing Company, Urban Infrastructure and Environmental Products Company, and High Performance Plastics Company. Prefabricated housing accounts for over half of Sekisui's sales.
1948	Carl Strandlund starts the Lustron Corporation, which sells about 2,500 of its all enameled-steel houses before closing in 1950.
1949	Designers Charles and Ray Eames finish their famous one-off home in California, using industrially-produced component parts, as part of the Case Study House program.
	Buckminster Fuller introduces his Wichita House, a lightweight, round, standardized aluminum structure. Only two are eventually built.

Year	Designer/ Influencer/ Major Achievement
1950	Jean Prouvé commissioned by the French government to create mass-produced housing. Twenty-five units are produced and installed in Meudon, France.
1953	Carl Koch designs the Techbuilt House, a wooden frame structure and panelized system.
1954	Australian architect Harry Seidler creates a prototype production house, a system of prefabricated columns, sections, and beams to allow for extreme flexibility in floor plans.
1959	William Berkes, a graduate of the Harvard Graduate School of Design and disciple of Walter Gropius founds Deck House, a prefabricated residential building system.
1963	PanaHome Corp. engages in prefabricated detached housing, asset management, and home remodeling businesses in Japan. The company was founded in 1963 and is headquartered in Osaka, Japan.
1967	Misawa Mokuzai, the predecessor of Misawa Homes Co., Ltd., started business operations upon acquisition of the housing industry's first certification from the Ministry of Construction in 1962. Misawa Homes Co., Ltd. established in 1967. Misawa Homes announced as No.1 seller of homes in the prefabricated housing industry in Japan in1971. Misawa Homes attempts to produce homes that can be used by several generations over 100 years.
1967	Moshie Safdie's Habitat Montréal is built for the World Expo. 158 concrete modules stacked atop each other contained 18 different versions.
1968	Richard Rogers proposes his Zip-Up Enclosures, a series of standardized components that users could purchase to expand a living structure.
	Paul Rudolph is commissioned by the Amalgamated Lithographers of America to create more than 4,000 prefabricated living units (unrealized).
1971	Paul Rudolf 's modular housing project Oriental Masonic Gardens completed in New Haven, Connecticut.
1972	Kisho Kurokawa's Nakagin Capsule Tower in Tokyo is realized with living units that can be changed out over time.
1974	Zvi Hecker's Ramot Housing Complex in Jerusalem contains 720 polyhedric modules arranged in a beehive configuration.
1975	Toyota, the world's second-biggest auto maker behind General Motors, entered the housing business in 1975. Toyota's house making is based on the "Skeleton & Infill" approach. Now, Toyota homes are mass produced like cars. About 85% of the work on the metal-frame cubicles is finished at the plant. The prefabricated cubicles, made to order for the customer, are stacked like toy blocks with a huge crane and topped with a roof in just six hours.
1980	The National Mobile Home Construction and Safety Act is renamed the National Manufactured Housing Construction and Safety Act reflecting the difference between truly mobile recreational vehicles and more permanently sited manufactured homes.
1985	Deborah Burke creates the Single Wide and Double Wide, two modular house designs, for developer Harvey Gerber.

Year	Designer/ Influencer/ Major Achievement
1993	Mark and Peter Anderson develop their first balloon-frame panel house on Fox Island, Washington.
1995	Shigeru Ban completes Furniture House in Japan, which uses factory-finished and site-installed floor-to-ceiling shelving as structural support for the roof.
	Wes Jones uses standard shipping containers as the basis for his Technological Cabins series.
1996	Mass-market retailer IKEA introduces its more traditional style Bo Klok house in Sweden.
1997	KFN Systems in Austria completes Two-Family House, a timber-framed house based on a modular system created with a group of traditional carpenters.
1998	Austrian-based firm KFN introduces its prefabricated module SU-SI, which is trucked to the site and erected on piers.
2000	Global Peace Containers, a non-profit organization that converts retired shipping containers into housing and community buildings, completes a school in Jamaica.
2001	Sean Godsell creates Future Shack, an emergency housing prototype built from a discarded shipping container.
	David Hertz creates his Tilt-Up Slab House in Venice, California, utilizing precast concrete panels.
	Rocio Romero offers the LV Home, a Galvalume-clad rectangular, flat-roof, glass walled home as a partial kit home.
2002	Allison Arieff and Bryan Burkhart publish Prefab, the first survey of contemporary prefabricated houses and their historical antecedents.
2003	The architectural firm of LOT-EK completes its prototype for a Modular Dwelling Unit, a shipping container converted into a home featuring extendable and retractable modules that increase usable interior square footage.
	Alchemy Architects completes its first weeHouse, a one-room prefabricated modular cabin in rural Wisconsin.
	Dwell magazine launches its prefab house competition. A design by Resolution: 4 Architecture is selected for construction at a North Carolina site.
2004	Michelle Kaufmann's Glidehouse, a modular prefab home, debuts at Sunset magazine's Celebration Weekend in Menlo Park, California, and enters production.
	Charles Lazor completes his prototype FlatPak, a panelized prefabricated system, in Minneapolis. In 2005 the FlatPak is offered by Empyrean International LLC.
2005	Marmol Radziner debut Desert House, a prototype for a steel-welded frame modular prefab system of living, shade, and deck modules. A factory is established in Los Angeles for production controlled by the architectural firm.
	Michelle Kaufmann debuts the Sunset Breezehouse, a modular prefabricated home featuring a series of garden spaces and a choice of roof type, and enters production.

Table 2.2. Prefabricated Housing Development Timeline from 1906-2007 (Page 5 of 5)

Year	Designer/ Influencer/ Major Achievement
2005	Pinc House based in Sweden adds Black Barn to its home offerings. Based on the Viking longhouse, the pitched-roof structure uses a panelized prefabricated system.
	Steven Holl Architects completes Turbulence House, a stress-skin metal panel house in New Mexico, which utilizes computer-controlled cutting technology to produce 24 unique panels.
2006	LivingHome, created by Ray Kappe, a leading modern architect and the founder of the Southern California Institute of Architecture (Sci-Arc). All LivingHomes products are designed to achieve at least a LEED for Home "silver" accreditation. The first model home in Santa Monica, CA was awarded a LEED Platinum rating in August 2006.
2007	Jeriko House is a prefab modular system that made by patented aluminum framing which is light, rigid and impervious to rust, mold and weather. The system allows unlimited configuration choices and encourages design collaboration and innovation.

2.2.1 Sears & Roebuck Mail-Order Kit Houses, 1908 – 1940. From 1908–1940, Sears, Roebuck and Company sold about 75,000 homes through their mail-order Modern Homes program. Over that time Sears designed 447 different housing styles, and customers could choose a house to suit their individual tastes and budgets (Thornton, 2004). Although Sears was not an innovative home designer, Sears mail-order kit houses can be viewed as the first customer-tailored mass product in housing industry. Sears provided a house plan catalogue (Figure 2.2) with the added advantage of modifying houses and hardware according to buyer tastes, and shipped the appropriate precut and fitted materials to the customer's site.

Figure 2.2. Sears Modern Homes Catalog, 1924

12

Sears mail-order houses began to decline in the late 1920s. Automobiles and improved roads made it easy for rural residents to travel long distances to shop. To adjust to this development, Sears began establishing retail stores in 1925 and 1926, respectively, to tap the growing business in larger towns. Sears catalog sales had dropped to only 54 percent of the firm's total business by 1930. Finally, the Sears Modern Homes department closed in 1940 (Thornton, 2004).

2.2.2 Buckminster Fuller – Dymaxion Deployment Unit Project, 1927. Buckminster Fuller had first envisioned the Dymaxion House in 1927 (Figure 2.3). He saw it as a house that touched the earth lightly through an efficacy in structure and materials. He created a list of criteria that all houses should meet: mass-production, strength, low maintenance, and light weight. Aluminum was the material of choice for Buckminster's house, although expensive at the time, which required diligent efficacy in its use. Aluminum is recyclable, does not require painting, is the second most common element on Earth, and has a long life span.

Figure 2.3. Dymaxion House

Figure 2.4. Wichita House

13

Commissioned by the Army for field housing, hundreds of these units were shipped to the Persian Gulf during World War II. Fuller's continued research into this production method led to the development of the Wichita House in 1949 (Figure 2.4).

In 1945, Fuller formed an alliance with Beech Aircraft Company of Wichita, Kansas to produce the Wichita Dymaxion House (Wichita House). World War II had just ended with thousands of soldiers returning home and needing housing and jobs alongside wartime plant workers. The Dymaxion House was seen as a way to retool the plane factories and provide housing (Baldwin, 1996).

The aerodynamic shape of the space and the mast work to bring the interior temperature down 15 degrees below exterior temperatures through airflow alone, without air conditioning. Fuller's concern with creating an inexpensive and efficient solution to the problem of housing led him to develop the Wichita House so that it could be erected by a single person.

However, all of these systems were purely theoretical and never were fully tested in factory or in the field. Once the Wichita house was inhabited, the rudder on the top of the mast was immobilized due to its noise. The natural cooling/heating system no longer worked. The owners installed conventional systems, which the Dymaxion was not designed for and the house produced large heating and cooling loads.

Ultimately, only two Dymaxion kits were ever built. These two kits were cannibalized into one house in Wichita, Kansas. It was plagued by problems due to improper installation. It leaked, as it was designed to, because the interior perimeter gutter had not been installed, which was designed to collect the water from the roof panels. The contractor also installed the aluminum panels with steel fasteners thus causing electrolytic corrosion.

The owner died in the 1970's and his children moved away leaving the Dymaxion to the raccoons. The Dymaxion has since been refurbished and relocated to the Ford Foundation in Dearborn, Michigan.

2.2.3 Gropius & Wachsmann – Packaged House, 1941-1952.   Konrad Wachsmann and Walter Gropius designed and produced what they called the "Packaged House" (Figure 2.5), a panelized system for the construction of houses between 1941 and 1952. They founded a company called the General Panel Corporation to manufacture the system, and received tremendous publicity, government support and investor backing. Gropius, in particular was an architect of international renown, founding director of the Bauhaus in Dessau, Germany and dean of the Harvard Graduate School of Design after coming to the United States in 1937. Wachsmann was well known in Germany for building Albert Einstein's house in Potsdam in 1929, and for his book Building the Wooden House of 1930. Even with such renowned and strong backing, their company failed after producing only 500 houses. The design of their system is inspiring, while the fate of their company is a precautionary tale. Architects interested in prefabricated housing typically look to the automotive industry as a model for increasing quality and decreasing cost through mass production (Wachsmann, 1995).

The important idea of the Packaged House was a system rather than a house, at least in its initial conceptualization. It was designed to be customized by other architects or owners for multiple configurations. The Packaged House system was based on load-bearing wood framed panels, which could be used without any other structure. Pieces were interchangeable, which allowed ease of assembly, with no measuring required. Five workers could assemble the house in one day. Every wall panel was the same size with the module of 3'-4" wide by 8'-0" high (Figure 2.6). But as Gilbert Herbert, author of The Dream of the

Factory-Made House points out, while the design was an open system, the construction was not, and this became one of its downfalls as we will see later (Herbert: 1984).

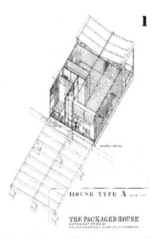

Figure 2.5. Packaged House, Standard House Type A, 1942

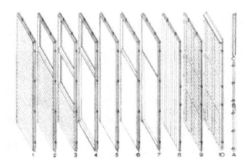

Figure 2.6. Interchangeable Walls for Packaged House System

The key to the system was its steel connector, which was invented by Wachsmann. All connectors in all panels were the same, which made factory production simpler (Wachsmann, 1961). It also made installation easier, to be accomplished by unskilled workers. It was universal; only one type of connector was used throughout the entire house, and it worked either horizontally or vertically.

16

One reason often cited for the failure of the company was Wachsmann's obsessive perfectionism with regard to the design of the factory and the machines (Herbert, 1984). This took several years, also hampered by problems with capital flow due to organization in a weak company. Unfortunately the delays meant that the company missed a critical moment in post-war demand and government funding of their loans expired.

The fact that the system was cohesive in design, that everything was made in the factory, is also given as cause for its failure. It did not allow for inclusion of pieces from other sources, which could have speeded initial production. Also, since the module did not conform to standard plywood sizes, there was an inherent waste, and loss of ability to generate competition between subcontractors. The module was based on the metric system, rather than the American inch-foot system, which made the whole system even more separate from the American building industry. The company was never able to achieve enough volume to bring the price down. At the same time, site-built house developers became sufficiently mechanized and large enough to take advantage of economies of scale to drive down costs to reach a wider market.

As Herbert notes, the house is the most personal commodity for Americans. Gropius and Wachsmann had little knowledge of marketing and positioning their product within the American consumer society (Herbert, 1984). American site built developers could provide much greater variety in their products, and incorporated mechanized means of building to bring their costs down. The lessons to be learned from the General Panel Corporation System House's failure are the importance of awareness of public perception and the possibility of incorporating other manufactured products.

2.2.4 Jean Prouve – Meudon Housing, 1949-1950. Jean Prouve, an industrial and architectural designer, was commissioned by the French Minister of reconstruction and planning to design a new kind of mass produced housing in 1949 that would cost no more than the cheapest existing housing. These houses consisted of a minimal structural system of folded sheet steel and wood panels. The French government ordered twenty five units. In 1950, the houses were installed at an experimental housing project in Meudon (Figure 2.7) (Arieff and Burkhart, 2002). The initial schemes show fourteen variations on two unit types.

Figure 2.7. Meudon House type 1, 1950

Construction Sequence:

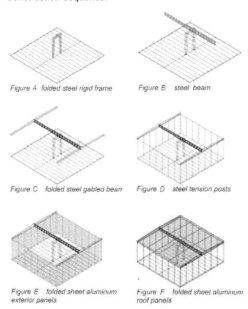

Figure A folded steel rigid frame   Figure B steel beam

Figure C folded steel gabled bean   Figure D steel tension posts

Figure E folded sheet aluminum   Figure F folded sheet aluminum
exterior panels   roof panels

Figure 2.8. Construction Sequence Diagram of Meudon House

Figure 2.9. Tropical House (Prototype after Meudon House), 1951

Prouve devised a jointed steel structure that could be erected without scaffolding, a technique he used on a number of occasions, notably with Pierre Jeanneret and Charlotte Perriand for housing, hospital and other buildings in 1939. The structural members, made of folded steel, consist of one or more lateral portals, in the form of a truncated "A", pin-jointed to a central longitudinal ridge member with slightly angled gable beams bolted laterally at both ends (Figure 2.8). The roof panels are arranged laterally at one end and over the prefabricated panels at the other. The houses themselves are raised off the ground on tapering steel joists supported on piloti. The positioning of the partitions inside is relatively flexible, limited only by the structural module of the external panels (Figure 9). However, the incorporation of masonry for the basements and even in the interiors of some houses hides the evidence that the homes were prefabricated in a shop. Prouve's intention in the original design was to have the houses supported by steel pilotis, with a basement enclosed with aluminum panels.

The houses on the Meudon estate were sold to fairly wealthy people. Although some modifications and changes have been made to individual houses over the last 35 years, the estate is still in very good condition. Prouve demonstrated that Meudon houses could be produced competitively, which suggested that they might be manufactured on a large scale. However, the government never again took up the design and no more houses were produced.

The Meudon houses show a unique structural system that can be constructed using a number of different materials. His design for these houses proved that light and industrially fabricated housing systems could be mass produced with customizable elements for a cost competitive with conventional building methods (Huber and Steinegger, 1971). In the United States, housing was already being constructed out of mass produced elements, when Prouve was developing his plans for Meudon. Unlike Prouve's work, most American examples of the period (perhaps with the exception of Buckminster Fuller) were more concerned with affordability through standardized parts rather than innovative structures.

2.2.5 Lustron Houses, 1949.   Lustron was the vision of Carl Standlund, a vice president of a porcelain enamel company that created baked on enamel-steel panels for gas stations. His goal was to produce a cheap, efficient, mass-produced house (Figure 2.10) that would address the growing housing shortage at the end of World War II. His solution, intended as an inexpensive factory-production system, was designed to meet high-volume demand for housing. Consequently, he won support from the Federal government in the form of huge low-interest loans, and within a year had secured enough capital to finance a twenty-three acre factory and begin a nationwide marketing campaign. The actual machinery for producing the houses was tremendously elaborate and expensive, and produced and exorbitant number of parts (3,000) that had to be skillfully assembled on-site (Fetters, 2002).

Figure 2.10. Lustron House

The problems with the Lustron system were manyfold, but can be traced back to the initial inception of the design. From the very beginning Standlund set unreasonable constraints. That the home should be entirely made of metal was problematic from start to finish. That assembly required skilled labor was bound to cause mistakes, loss of time, and ultimately money. The singular floor plan and arrangement of pre-determined bookshelves, cabinets, and vanity mirrors was overbearing and not able to change as needs might change. And above all, given the huge amount of capital invested in building machines and production lines, that the actual manufacturing process was so rigid and essentially impossible to change seems unreasonable. A contradiction in manufacturing was miraculously achieved: the product, while hyper-manufactured, was totally un-designed. As a result, the failure significantly undermined any confidence that the American public, and more especially lending firms, had in pre-fabricated building systems.

Reviewing the industrial failures from prefab housing history, there are two significant examples from the late 1940s: the Lustron All Metal Dream Home, and the General Panel Packaged Home. Lustron was financed and conceived of by an industrialist, while General Panel was the vision of two great architects, Walter Gropius and Konrad Wachsmann; neither of the efforts was successful architecturally or financially. In both cases, all production took place in factories and the product was shipped as a "kit of parts" for rapid on-site erection. The Lustron had a totally fixed plan and offered no flexibility, while

General Panel was intended to be a highly adaptable, limitless means of creating space with a system of universal connectors (Wachsmann, 1961). Both of them tried to solve the housing problems by proposing over-simplified design solutions. Lustron system tried to reduce the problem of the house by using a single material, and General Panel Packaged Home tried to reduce the problem to a single connection.

In the end, the Lustron Corporation only produced some 2,500 houses, far from their original goal of manufacturing 25,000 homes per year (Fetters, 2002). For General Panel Corporation, the total number of their finished products did not exceed between 150 and 200, which caused the factory to be shut down. The dream of the factory-built house was actually a financial nightmare, because the profits needed a function of a sustained output volume (Herbert, 1984). Volume is a big challenge to all factory-built house creators, because their business can not survive if the initial cost for the factory can not be paid off by eventual housing production output. Unlike the automobile industry, people do not need to replace a new house for every five years, and the price of a prefab house is far beyond the price of a new car.

2.2.6   Bertrand Goldberg – Prefabricated Housing (Unishelter), 1953.

The designer of Marina City, Chicago-based famous architect Bertrand Goldberg had produced a great number of industrialized products after his return from Bauhaus until the end of World War II. He was intent on combining the concepts of economics in German architectural design he learned from Bauhaus with the American economics of mass production. In 1936-1937, he designed five prefabricated houses, which were sold in one day. In 1939, he received a commission for fifty houses made with plywood panels prefabricated in a factory, and established a production system that was capable of delivering ten bathrooms per day (Ragon, 1985).

The Modernistic Unishelter was manufactured in stress skin plywood construction for the U. S. Army and shipped to Alaska. These housing units were able to be packed with material and equipment during the shipping, and therefore served a double purpose as a container during shipping and a house after arrival. The basic unit of Unishelter combines service core, living room, and bedroom. The secondary unit is slightly shorter in length for another extra two bedrooms (Figure 2.11). These units were manufactured separately and could be purchased separately. The Unishelter units could also be combined in many ways to form shopping units, clinics, or as schools besides serving as housing.

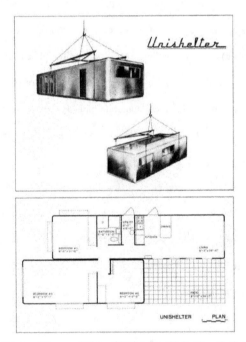

Figure 2.11. Primary View and Plan of Unishelter Unit, Hegwisch, Illinois, 1953

2.3   Types of Prefabricated Housing System

Generally, there are three different types of prefabricated housing systems: fully modular, sectional, and component. In Figure 2.12, it describes the basic modular types,

23

features, and examples for each system. The complexity of on-site assembling and shipping limitation may be related with the scale of basic modular elements. This analysis will help us to understand the strength and potential of each system, and provide opportunities for customization and spatial adaptability.

Basic Module	Feature	Precedence	Current Example	
Fully Modular	• As 3D modules (like boxes) • Simple connections to the foundation • Size of the modular unit is restricted by highway or shipping constraints	 Habitat '67	 weeHouse	 Dwell Home
Sectional	• Sectional modules for transport easily • It has some potentials for digital fabrication	 Double-Wide	kitHAUS	 ESG Pavilion
Component	• Factory-made components to reduce the on-site labor • Allows flexible building shapes • Includes Panelized, Precut, Kit-of-parts system	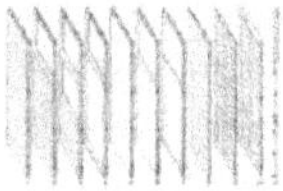 Packaged House	 FlatPak	 iT House

Figure 2.12. System Types by Basic Modular Element

2.3.1 Fully Modular System. All the components of a single housing unit are entirely made, assembled and finished at the plant; as three-dimensional modules (like boxes) requiring only simple connections to the foundations and main service conduits once at the site. The size of the modular unit is restricted by highway law or shipping constraints. Some examples are like Habitat '67 in Montreal, Canada by Moshe Safdie and Nakagin Capsule Tower in Tokyo, Japan by Kisho Kurokawa.

2.3.2 Sectional System. Small and easy to transport sectional modules, but need a complementary component or process once they reach the site. Double-wide trailer is the typical example found along the North American highways. Sectional modular system has some potential for implementing with digital fabrication technology. ESG Pavilion by

24

graduate students in Swiss Federal Institute of Technology (ETH) is an example of creating sectional modules.

2.3.3 Component System.   By definition, a component system may be a panelized, precut, or kit-of-parts system. All building components are pre-engineered and designed to be assembled in a variety of ways. Components are sized for convenient handing or according to shipping constraints. The smaller of components may take longer time to be assembled on site, but that allows more flexibility for creating building variations. SIP (Structural Insulated Panels), the new generation building materials, is high-performance prefabricated building panel for walls, roofs and floors in residential and commercial buildings.

2.4   Mass Customization: Concept and Application

2.4.1 Concept of Mass Customization.   At the beginning of the twentieth century, industrialized economies were focused on mass production, mass distribution, mass marketing and mass media. Presently, a combination of advances in information and digital technology is making it increasingly possible to rapidly respond to consumers with customized products at mass-production prices.   The fundamental premise of mass customization is to no longer manufacture products "blindly" according to a predicted demand, but instead allow production to be directly driven by actual orders (Schodek, 2004).

The term "mass customization" was coined by Stan Davis in his book Future Perfect but the term was popularized by Joseph Pine in his book Mass Customization: The New Frontier in Business Competition in 1993 (Schodek, 2004). Mass customization has different implications for different products and in different sectors. There are also different methods and strategies to achieve it (Crayton, 2001). Some products can be tailored or customized directly by the consumer; other products may only have limited degree of customization at the retail outlet or dealer.

25

The key to cost effective customization is modularization and configuration (Crayton, 2001). One of the key ideas and strategies to achieve mass customization is modularization. Products are "decomposed" into modular components or subsystems that can be recombined to more nearly satisfy customer's needs. The modularization approach is very close to the spirit of prefabricated housing, and this model can be viewed as a revitalization of mass production housing. The configuration systems present the choices to consumers and determine what goes with what. Figure 2.13 illustrates that by utilizing web technology, the configuration systems can be represented as a design interface to convert customer's input to final product's configuration, and the on-demand production by the assembly line combine standard modules together (Schodek et al., 2004).

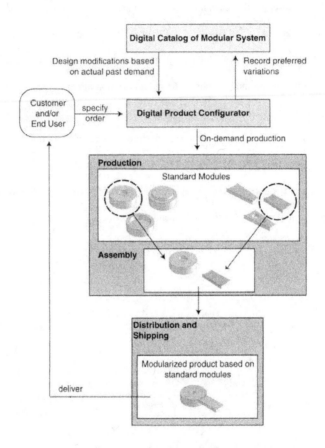

Figure 2.13. Schematic Process Diagram of Mass Customization

To summarize the differences between mass production and mass customization, Table 2.3 Mass Customization contrasted with Mass Production is quoted from Joseph Pine's Mass Customization: The New Frontier in Business Competition.

Table 2.3. Mass Customization Contrasted with Mass Production

	Mass Production	Mass Customization
Focus	Efficiency through stability and control	Variety and customization through flexibility and quick responsiveness
Goal	Developing, producing, marketing, and delivering goods and services at prices low enough that nearly everyone can afford them	Developing, producing, marketing, and delivering affordable goods and services with enough variety and customization that nearly everyone finds exactly what they want
Key Features	Stable demand	Fragmented demand
	Large, homogeneous markets	Heterogeneous niches
	Low-cost, consistent quality, standardized goods and services	Low-cost, high-quality, customized goods and services
	Long product development cycles	Short product development cycles
	Long product life cycles	Short product life cycles

In terms of the implementation of AEC (Architecture, Engineering and Construction) industry, the fabrication of modules is highly dependent on the repetitive mass production of identical parts; however, the independency of these parts allows for flexibility as they are easily interchangeable with other similar parts. This creates opportunities for the production of a diverse range of end products; end users have the ability to easily personalize products that were initially fabricated from an array of mass produced identical parts. This method for achieving variety depends greatly on decisions made by designers at the early stages of the conceptual design process (Barrow, Arayedh & Kumar, 2006). This requires breaking down products into smaller parts and defining the relationships between the parts. These strategies for mass customization as outlined by Joseph Pine (1993) are as follows:

28

Component-sharing Modularity: Common components used in the design of a product. Products are uniquely designed around a base unit of common components. Example: elevators.

Component-swapping Modularity: Ability to switch options on a standard product. Modules are selected from a list of options to be added to a base product. Example: personal computers.

Cut-to-fit Modularity: Alters the dimensions of a module before combining it with other modules. Used where products have unique dimensions such as length, width, or height. Example: eyeglasses.

Mix Modularity: Also similar to component swapping, but is distinguished by the fact that when combined, the modules lose their unique identity. Example: house paint.

Bus Modularity: Ability to add a module to an existing series, when one or more modules are added to an existing base. Example: tracking lighting.

Sectional Modularity: Similar to component swapping, but focuses on arranging standard modules in a unique pattern. Example: Legos.

Figure 2.14 provides iconic illustrations of each of the six different kinds of modularity, which are all discussed above with examples.

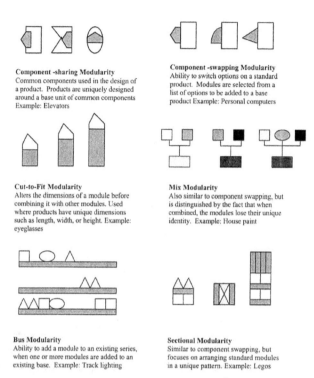

Component-sharing Modularity
Common components used in the design of a product. Products are uniquely designed around a base unit of common components Example: Elevators

Component-swapping Modularity
Ability to switch options on a standard product. Modules are selected from a list of options to be added to a base product Example: Personal computers

Cut-to-Fit Modularity
Alters the dimensions of a module before combining it with other modules. Used where products have unique dimensions such as length, width, or height. Example: eyeglasses

Mix Modularity
Also similar to component swapping, but is distinguished by the fact that when combined, the modules lose their unique identity. Example: House paint

Bus Modularity
Ability to add a module to an existing series, when one or more modules are added to an existing base. Example: Track lighting

Sectional Modularity
Similar to component swapping, but focuses on arranging standard modules in a unique pattern. Example: Legos

Figure 2.14. Modularity Types[5]

2.4.2 Current Applications. Compared to approaches of mass customization in industrial product design, like Nike-iD[6] series products (Figure 2.15), there are more challenges to apply this model to architecture. From design to construction, a new building is a complex process involving a number of independent parties. There is usually no one party that is expert in all areas, and this industry-specific fragmentation is a major obstacle to mass customization. However, there are a few companies that have successfully adopted mass customization concept to their individual products. A good example as a case study is E-

[5] Ulrich, Karl and Tung K. "Fundamentals of Product Modularity", Proceedings of the 1991 ASME Winter Annual Meeting Symposium on Issue in Design/Manufacturing Integration, Atlanta, 1991.

skylight.com supported by Architectural Skylight Company (ASC)[7] (Figure 2.16) is a good example as a case study.

Architectural Skylight Company uses an object-oriented design approach to the design and manufacturing of custom skylights. This system supplements AutoCAD with several plug-ins, including third-party software and programs developed by ASC. The website interface provides a step by step customizing process to generate a final design model, and the virtual model is used directly with computer numerical control (CNC) for manufacturing of frame members and for the CNC cutting of custom glass sheets.

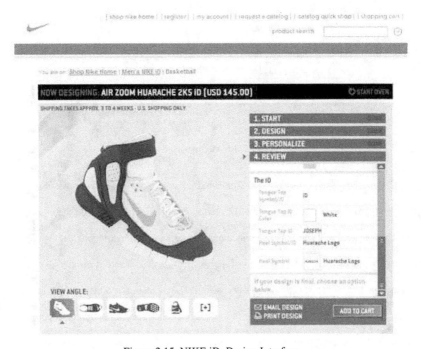

Figure 2.15. NIKE iD, Design Interface

---

[6] NIKE iD, http://niketown.nike.com/niketown/catalog/category.jsp?categoryId=53595
[7] E-Skylight.com, http://www.e-skylight.com

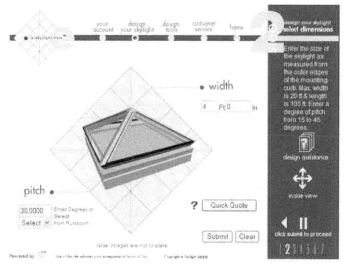

Figure 2.16. Design Interface of e-skylight.com

2.5   Current Researches in Mass Customizing Houses

Based on the thesis topic, this section reviews the related academic researches including the international conference papers from last decade and two significant research groups of digital technology and housing in MIT and McGill University. The conference papers have been selected with the theme of mass customization, internet-based computer-aided design, and consumer participative design with directly or indirectly apply to the prefabricated modular housing.

Table 2.4 indicates the current researches in modular housing with digital application. There is no clear beginning date of this research topic, because prefabricated modular housing has been implemented in the twentieth century without any digital application. During the early years of twenty-first century, many researchers focused on the internet information technology to strengthen the traditional collaborative design. With the popularity and development of digital fabrication and computer programming, many research papers in 2006 discussed these digital technologies with the case of prefabricated modular houses.

Next two sections introduces the more in-depth researches carried by some research institutes.

Table 2.4. Current Researches in Modular Houses with Digital Applications (Page 1 of 3)

Researcher(s)	Year	Title of the research paper	Research Content
Combes, L. & Bellomio, A.	1999	Creativity and Modularity in Architecture (AVOCAAD)	Modular thinking & standardized free form by CAD
Sarid, A. & Oxman, R.	2000	The Web as a Knowledge Representational Media for Architectural Precedents (eCAADe)	Explore the potential of web media for architecture
Chien, S. & Shih, S.	2001	Design through Information Filtering: A Search Driven Approach for Developing a Layperson's CAAD Environment (CAAD Futures)	Design knowledge encapsulation, CAAD for non-designers
Donath, D. & González, L.	2001	Integrated Planning Support System for Low-Income Housing (SIGRADI)	Participative planning with IT tools
Stouffs, R., Tunçer, B. & Sariyildiz, S.	2001	The Customer is King: Web-Based Custom Design in Residential Developments (CAADRIA)	Web-based client participative design through a series of games
Rizal, H. & Rafi, A.	2002	The Impact of Internet Enabled Computer Aided Design (iCAD) in Construction Industry (CAADRIA)	Digital communication application in construction industry
Zhou, Q., Krawczyk, R. & Schipporeit, G.	2002	From Cad to iAD: A Working Model of the Internet-Based Engineering Consulting in Architecture (CAADRIA)	Working model to integrate internet applications in the AEC industry
Zhou, Q., Krawczyk, R. & Schipporeit, G.	2002	From Cad to iAD: A Web-based Steel Consulting of Steel Construction in Architecture (eCAADe)	Web-based engineering consulting system
Duarte, J.	2003	A Discursive Grammar for Customizing Mass Housing (eCAADe)	Mass customizing housing by shape grammars
Fröst, P.	2003	A Real Time 3D Environment for Collaborative Design (CAAD Future)	Customer-driven collaborative design in a real time 3D computer game

Table 2.4. Current Researches in Modular Houses with Digital Applications (Page 2 of 3)

Researcher(s)	Year	Title of the research paper	Research Content
Fure, A. & Daubmann, K.	2003	Mass-Crafting Numerical Instructions for Construction (ACADIA)	Mass-customization in housing design by CAM
Zhou, Q.	2003	From Cad to iAD: A Prototype Simulation of the Internet-based Steel Construction Consulting for Architects (CAADRIA)	Internet-based consulting of structural system selection
Kim, J. & Jeong, Y.	2004	Collaborative CAAD: State-of-The-Art and The Future (CAADRIA)	Identify requirements of the design collaboration systems
Cavieres, A. & Quezada, M.	2005	Analysis of the possibilities offered by the application of parametric modeling technologies in the design processes shared between architects and industrial designers: The prefabricated house case (SIGRADI)	Case-based computer-aided design pedagogy
Barrow, L., Arayedh, S. & Kumar, S.	2006	Performance House 1: A CAD/CAM Modular House System (ACADIA)	Manufactured housing system by CAD/CAM
Botha, M. & Sass, L.	2006	The Instant House: Design and Digital Fabrication of Housing for Developing Environments (CAADRIA)	Design production with digital fabrication
Garber, R. & Robertson, N.	2006	The Pleated Cape: From the Mass-Standardization of Levittown to Mass Customization Today (ACADIA)	Mass-customization in housing design by CAM
Hecker, D.	2006	Dry-In House: A Mass Customized Affordable House for New Orleans (SIGRADI)	Mass customized affordable housing system by CNC
Huang, C., Krawczyk, R. & Schipporeit, G.	2006	Mass Customizing Prefabricated Modular Housing by Internet-Aided Design (CAADRIA)	Exploration of prefab housing and client's design input online
Huang, C., Krawczyk, R. & Schipporeit, G.	2006	Integrating Mass Customization with Prefabricated Housing (ASCAAD)	Conceptual model of mass customizing prefab housing
Huang, C. & Krawczyk, R.	2006	i_Prefab Home: Customizing Prefabricated Houses by Internet-Aided Design (eCAADe)	Web-based design advisory system for prefabricated housing

Table 2.4. Current Researches in Modular Houses with Digital Applications (Page 3 of 3)

Researcher(s)	Year	Title of the research paper	Research Content
Rahman, R. & Day, A.	2006	A Comparative Study of Digital and Traditional Tools for Participative Design (ASCAAD)	Comparison of traditional and computer-based participative design
Huang, C.	2007	A Choice Model of Mass Customized Modular Housing by Internet Aided Design (IMCM + PETO)	Customer participative advisory system
Huang, C. & Krawczyk, R.	2007	A Choice Model of Consumer Participatory Design for Modular Houses (eCAADe)	Questionnaire approach, design knowledge representation

2.5.1 MIT House_n Research Group. Most of people know that the MIT Media Laboratory occupies a leading position in the rapidly evolving new digital technologies. It was formed in 1980 by Professor Nicholas Negroponte and former MIT President Jerome Wiesner, growing out of the work of MIT's Architecture Machine Group, and cooperating with a range of other disciplines.

In the pioneering stage during 1960s, the most highly developed of design systems was URBAN5 (Figure 2.17) developed jointly by IBM and MIT. A system where a designer could easily converse with a computer was initiated. It attempted to develop software that would guide untrained people through all the complexities of contemporary building design. People could use this program to design their own house directly, without professional mediation. Ideally, if the building materials were intelligent and flexible enough, the house could shift into its new design automatically.

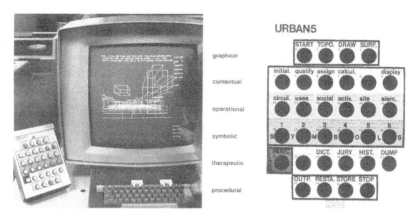

Figure 2.17. URBAN5 Model and its Function-key Buttons

The MIT House_n is a research group dealing with living space and digital technology. Founded in 1996 and directed by Kent Larson, it has a number of efforts underway to develop automated, web-based design tools called "preference engines" and "design engines". A preference engine engages the user in a dialogue to uncover needs, preferences, and values and the tradeoffs they are prepared to make. The system will approximate the dialog that a good architect may have with a client at the beginning of the design process. A design engine can be thought of as computation system that makes use of a set of rules encoded into a shape grammar that defines the architectural strategy of the designer. House_n researcher Lawrence Sass (Ph.D. thesis, MIT Department of Architecture, 2000) systematically extracted these rules from Palladio's text and the construction practices of the time to develop the basis for their codification into a shape grammar and Palladian design engine. This project extended the work of George Stiny (Professor, MIT Department of Architecture) and William J. Mitchell (Dean, MIT School of Architecture and Planning), who demonstrated that the design rules of Andrea Palladio can be inferred by analyzing other plans in the Renaissance publication, <u>The Four Books of Architecture</u>[8] . House_n researcher Jose Pinto Duarte (Ph.D. thesis, MIT Department of Architecture, 2001) explicitly encoded the design rules of the Portuguese architect Alvaro Siza for a mass housing project at Malagueira into a shape grammar and computer-based design engine that allows users to generate unique house designs and still remain the Siza's style (Larson, Tapia, and Duarte, 2001).

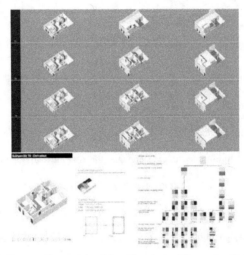

Figure 2.18. Jose P. Duarte, Customizing Mass Housing: a Discursive Grammar for Siza's Malagueira Houses

Although the end result of most thesis researches only resolves one dimension of the problems, MIT House_n research group is still leading the field in looking forward to how homes could be transformed over the next decade. In Jose P. Duarte's thesis, his design tool only dealt with the variation of interior space layout of Siza's Malagueira Houses (Figure 2.18), and not focused on the standardization of the factory-made components in order to

---

[8] MIT Open Source Building Alliance, http://architecture.mit.edu/~kll/OSBA/OSBA_october15-2002_white_paper.pdf

achieve mass customization. However, he put a lot of effort in researching a discursive grammar that can simulate the design rules of Alvaro Siza. Making use of Duarte's work, House_n researcher Xioayi Ma (Masters thesis, MIT Department of Architecture, 2002), has developed a web-based Universal Design Engine and decision-making tool for new kitchens with best-fit layout (Figure 2.19). This design algorithm leads the user through a series of steps to refine the design by exploring additional issues related to function, appliance specification, universal design attributes, and aesthetics.

Besides shape grammar and algorithmic design approach, Jennifer Beaudin[9] and the other students create apartment customization and kitchen design tools for laypeople to assist homeowners with translating their experience in use to the language of design as well as having a better communicate with builders. The tools are intended to situate the user within the design, to think about use issues, such as the distance between the sink and the Cooktop rather than the traditional critic-style tools.

---

[9] Beaudin, Jennifer S. From Personal Experience to Design: Externalizing the Homeowner's Needs Assessment Process. Master Thesis, Massachusetts Institute of Technology, 2003.

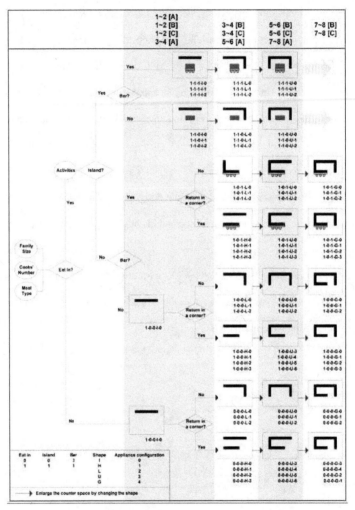

Figure 2.19. A Decision-Tree Example for the Parametric Design of Kitchens (Xioayi Ma, Masters Thesis, MIT Architecture, 2002)

Recently, Architect Kent Larson presented his radical vision that just might transform America's construction industry: Build houses the way we build computers. The house of the future should be more like a personal computer or a car. It should be affordable, built mostly in a factory, and with parts that are easy to repair or replace. You should be able to design your own home online, just as you can today with a Dell laptop or a Honda minivan. The key to making it happen is standardization and accessorization, learned from the lead of other industries (Stover, 2006). Every new house would have a structural frame, or "chassis," that would be expected to last 200 years or more. That chassis could be fitted with your choice of an endless selection of roofs, sidings, interior wall panels and electronics made by a variety of manufacturers. The home's parts would be replaceable as "infill" system which includes partition walls, wiring, plumbing and cabinets, etc.

2.5.2 Dr. Noguchi's Mass Custom Home Model.   Buying a new home is a significant investment usually undertaken only once or twice during the lifetime in Japan. Therefore, consumers are very careful when buying a house, because it must satisfy their personal requirements in view of the demographical changes of today's society. In other words, today's consumers are no longer satisfied with monotonous mass products, even though the products are reliable enough in terms of the product quality. Houses must be responsive to people's individual needs, especially when they are relatively costly.

Massa Noguchi proposed "A Choice Model for Mass Customization of Lower-Cost and Higher-Performance Housing" in his Ph.D. Thesis from McGill University, Montreal. The Ph.D. Program in McGill University is famous for their housing research and publications. Noguchi's thesis is based on the learning experience from Japan's prefabricated housing industry. In his research, he categorized today's housing design approaches in North America into three types: production, semi-custom, and custom. Production design approach allows homebuilders to produce "ready-built" homes for high volume construction. The total time to build a standardized house is much shorter than for a one-of-a-kind design since construction staffs are familiar with the plans. Semi-custom design approach combines the characteristics of ready-built and custom-built homes. Usually, customers begin with an existing floor plan, and they have enhanced opportunity to modify and customize the interior and exterior finishes of their new home. Semi-custom design approach is the current method to balance the standardization level (higher than custom home) and customization level (higher than ready-built home).

In Noguchi's model, the "mass custom" design approach optimized both standardization level and customization level (Figure 2.20). He quoted the general methods of mass customizing products and services that were systematized by Joseph B. Pine II. A house consists of many components, which can be considered as "products", while design, construction and marketing are usually regarded as "services". To generate a housing development, these two aspects are involved with housing materials and systems as the products and the design and construction of these homes as the services. Thus, the 'mass customization system' can be formulated conceptually as follows:

$$MC = f (PS)$$

Where "MC" denotes a mass customization system itself and "f" means factor; "P" is the product sub-system that helps housing suppliers to mass-produce housing components, while "S" is the service sub-system that involves the interaction with users (or buyers) which helps them to customize an end product (Noguchi, 2003).

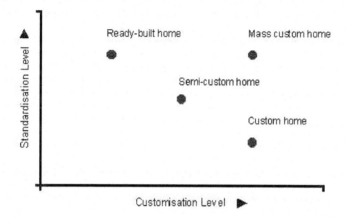

Figure 2.20. Standardization – Customization Relationship by Housing Type

Furthermore, in customizing products, "user participation" is vital, and housing suppliers need to offer design support communication services to their clients in locations with design-consulting staffs and appropriate communication tools to facilitate user's choices. These fundamental 'design-service' factors can be also integrated into a comprehensive model:

$$S = f\ (l, p, t)$$

In this model, the "service" sub-system is denoted by "S", and is supported by the existence of the location (l), personnel (p), and tool (t) factors (Noguchi, 2003). An important part of mass customization is that the user directly determines the configuration of their homes from choices given as client input during the design stage. This could not be achieved without the standardization of housing components. These are arranged in a visually attractive way in a component selection catalogue to enable clients to easily choose from the many options. The "product" sub-system (P), which can be explained by the following conceptual model:

$$P = f\ (v, e, i, o)$$

42

The volume (v) components are used to construct the structure of housing that determines the number and size of each room, while the exterior (e) and interior (i) components serve to co-ordinate both the decorative and the functional elements that customize a home. In addition, "o" denotes other optional equipment such as air conditioning, home security system, emergency call buttons, handrails, dishwashers and other electrical appliances (Noguchi, 2003).

The Mass Custom Home is a well-established guideline rather than a single-solution design prototype. The model illustrated the methodology of integrating mass customization concept into housing industry as well as the relationship of products and services. It may have the great potential to reform the current housing delivery system and contribute towards producing affordable mass-customized homes.

2.6   Internet Applications in Design Communication

Sears mail-order kit houses, from 1908 to 1940, can be viewed as the first customer-tailored mass product in the housing industry. Sears provided a house plan catalog with the added advantage of modifying houses and hardware according to buyer tastes, and shipped the appropriate precut and fitted materials to the customer's site. With today's technology, the internet is the perfect medium for the dissemination of domestic design. Many pattern book companies now have big websites offering thousands of house plans stored on databases searchable by type, style, square footage, average cost, number of bedrooms and so on (Davies, 2005). Some websites also provide the design tool for customizing wall materials, roof styles or interior finishes after the clients have selected the base model from a house plan catalog, examples like Original Home Plans (Figure 2.21) and Marmol Radziner Prefab (Figure 2.22).

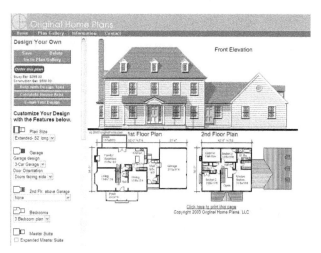

Figure 2.21. Website of Original Home Plans

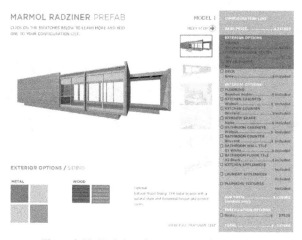

Figure 2.22. Website of Marmol Radziner Prefab

Misawa Homes Co., Ltd. established in 1967. Misawa Homes announced as No.1 seller of homes in the prefabricated housing industry in 1971. Until today, Misawa Homes company is still one of the biggest home builders with prefabricated construction in Japan. Image Search Lab, the web-based house plan searching engine shares their stock design plans by using the method of favorable impression. With different categories sorted by exterior, interior, square footage, entrance orientation, etc., the user can pin down the images by visual preference on the scale from the high to low (Figure 2.23). At the end, the interface will generate the search results which include the matched house plans and photo images.

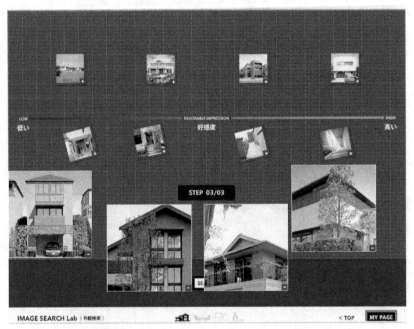

Figure 2.23. Image Search Interface of Misawa Home

Another example of internet-aided architectural design is etekt.com, an on-line electronic architecture studio. Founded in 1999, Etekt is a web-based platform for architects to present and sell innovative, high quality pre-designed single family home plans to private clients and home-builders[10]. Responding to clients' demand for increasingly customized house designs, and to architects' need for an expanded target audience, etekt.com restructure the conventional relationship between architects and clients. The use of technology for

---

[10] Etekt FAQ for clients, http://www.etekt.com/?L0=info&L1=faq_client&ID=

matchmaking between architects and clients has its parallels in computer dating services (Howler, 2001). Architects post their home designs in the marketplace. Clients seeking pre-designed home plans create a program defining their needs and desires and Etekt automatically guides them to an appropriate design or alternative options following by the percentage of match (Figure 2.24). Co-creation and Communication tools will be the future features of the marketplace, which allow the client to further customize his/her future home via an online dialogue with the respective architects and designers. Once the architect and client have reached an agreement, contact information is released. The architect adjusts the base plan to meet the client's requirements and finalizes the construction documents and specifications. The architect bills the client through Etekt website. Clients can use Etekt Billing and Payment services to send payments to respective firms. The client is asked to rate the architect's performance in an on-line questionnaire. Ratings will appear on the architect's profile for view by future Etekt clients. Much like readily available stock house plans, construction proceeds in a conventional process in which the client selects a local builder or contractor. Etekt employs existing technologies to capitalize on existing market trends and offers high-end design and a high-tech interface with low-tech implementation (Howeler, 2001).

Etekt makes it possible for architectural design and planning professionals to tap into mass-housing markets inaccessible to them today and for clients to find the appropriate designer and home design. In addition, Etekt minimizes the time and effort architects normally invest in acquiring new clients. Etekt unifies small architectural design and planning firms into a new market force, empowering each company to take full advantage of the U.S., and in the future, global residential home markets as a source for increased business.

Figure 2.24. Etekt.com, On-line Electronic Architecture Studio

The overall concept of web-based architectural design marketplace and on-line studio is innovative in this digital age. So far, etekt.com is only on the conceptual stage, because most design proposals from its database appear to be un-built projects and no realistic testimony from the clients to convince people that the marketplace is trustable. However, the methods of defining client's needs and desires are critical and can be discussed with further investigation.

Besides web-based architectural design marketplace, most common internet applications in architectural firms are email communication (ex. Microsoft® Outlook) and online project management and collaboration (ex. Autodesk® Buzzsaw). Many architects, engineers, real-estate professionals, and contractors are realizing the benefits of online project collaboration. These benefits include: improved project communication, shortened project cycles, increased ownership and accountability, improved record keeping, archival of information and continuous access to key project information in a secure and centralized location.

The Autodesk Buzzsaw software (Figure 2.25) can be run within the Microsoft Internet Explorer browser or as a stand-alone executable. The Autodesk Buzzsaw online project collaboration service has DWF (Drawing Web Format) at its heart, as good as viewing DWG files without having AutoCAD software installed[11]. Buzzsaw integrates with the Autodesk DWF Viewer and Autodesk Design Review. Buzzsaw automatically generates DWF files as part of the markup process. Review comments are stored as DWF files in relationship to original design data. Buzzsaw tracks all file versions. In addition, workflow forms like transmittals are linked to DWF files. The workflows are summarized via dashboards and reports. Design data is freely exchanged among project team members using DWF.

---

[11] By installing Volo® View, Volo® View Express, or Autodesk® Express Viewer software, users can view, mark up, measure, and print DWG, DWF™, DXF™, and raster format files from within the Autodesk Buzzsaw service without having AutoCAD® software installed. (www.autodesk.com/buzzsaw)

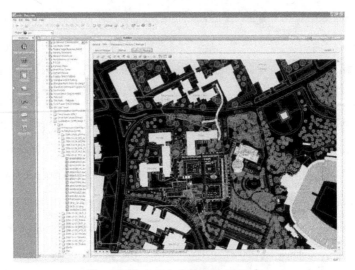

Figure 2.25. Autodesk Buzzsaw Interface

2.7   CAD Tools for Consumer Driven Home Design

There are many tools available at any local software store that have been developed to help homeowners design their home. 3D Home Architect® Home Design Deluxe 6 by Broderbund is an example of such software (Figure 2.26). With this software, you begin with a plan view of a design then drag and drop architectural elements from a menu and select finishes in a perspective view. One can subsequently view those designs in a 3D walkthrough and in high-quality renderings. This tool provides starting points, a library of parts to manipulate in design, and a variety of visualization techniques. It ties to certain manufacturers through off-line catalogs, provides a budgeting spreadsheet with material labor cost estimator, and the ability to print designs. The software, however, has some crippling faults. There is nothing to guide a user in selecting a starting point other than square footage and style, and nothing to differentiate how the spaces within those options might accommodate particular life-styles and activities. The visualizations offer little variety in lighting simulation, such as time of year and hour. The ties to industry are for appliances,

49

furnishings, and finishes, but not to the core architectural aspects of the home, for example walls, windows, hardware, or doors, nor to the labor force that would be involved in fabricating them. The budgeting feature is extremely suspect because pricing is tied to local markets and product selection, and is not fully automated. While final plans can be saved and printed, there is limited guidance on how to turn the final plans into a real home.

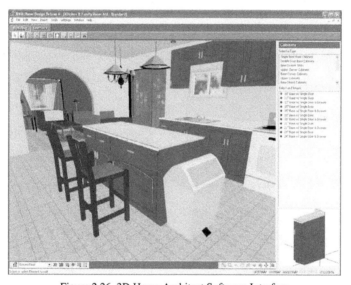

Figure 2.26. 3D Home Architect Software Interface

Similar to 3D Home Architect but more expensive and designed for home builders or advanced users, Chief Architect® can create custom plans and blueprints with a complete materials list (Figure 2.27). Chief Architect has been around for more than 10 years and has earned a solid reputation among its target group. It is aimed at the individual residential designer – it makes no provisions for larger projects that may require multiple users to work on the same project at the same time, and has limited applicability outside of house design market. The design is maintained in a single file that contains everything, i.e. foundations, floors, walls, and roofs. Chief Architect does allow multiple floor levels in the design file, with a maximum of ten floors. Because it is aimed for professionals, there are no guides to avoid impractical or even illegal solutions for less-experienced drafters or designers.

The House Wizard™ is the most interesting feature in Chief Architect that addresses the client's space requirement and translates the architectural program into a house floor plan. A series of dialog boxes requests basic program information input such as number of bedrooms and baths and whether you want a formal dining room. The program generates rectangular bubbles that represent the spaces. You arrange the spaces and fine-tune the room sizes and arrangement by dragging the rectangles. Once you're satisfied, you can convert the diagram into walls and the beginning of a 2D/ 3D house design model (Figure 2.28).

This unique feature seems intelligent in as much as it guides the users in creating a drawing with information input rather than starting from scratch. However, there are some missing parts in House Wizard. First of all, the "rectangular bubble diagrams" that contains itemized space blocks generated by House Wizard have no relationship to each other, and no guides for rational space adjacency. Unlike the traditional bubble diagram with linked lines in space planning, it definitely needs re-arrangement the size and location of each individual space by an experienced architect. It is not fully automated and no huge advance if compare with the method of beginning from a scratch. Second, the House Wizard routine tried to place appropriate doors where needed to provide access to rooms during the conversion process from rectangular bubbles to walls. However, it ends up with many inappropriate doors and only serves as a starting point for refinement. It neither makes intelligent choices about the door type and placement, or the relationships between two rooms with open or semi-open conditions.

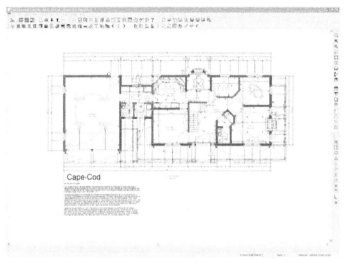

Figure 2.27. Chief Architect Software Interface

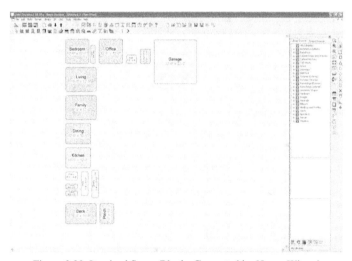

Figure 2.28. Itemized Space Blocks Generated by House Wizard

In addition to the complete residential design applications, there are some free wares online for consumers to make their home improvement. Bob Vila has a series of web-based home improvement tools to run in a web browser. One example is a tool to help an individual design a kitchen (Figure 2.29). It is also available from IKEA as a free download tool called Kitchen Planner (Figure 2.30). A user chooses items from a menu to be placed in a design, with controls for the viewpoint. The design space is fairly unconstrained, and there are no guides to help consumers to avoid impractical or even illegal solutions. The overall purpose is for kitchen vendors to sell their products easily via internet and provide a simple 3D viewer to simulate your future kitchen.

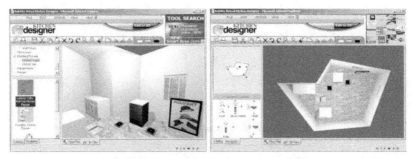

Figure 2.29. Kitchen Designer Tool from BobVila.com

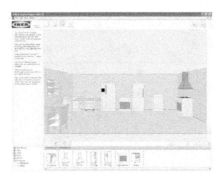

Figure 2.30. IKEA Kitchen Planner

2.8   Summary

Reviewing the history of prefab housing, there are a few examples that already proposed the concept of mass customization, like the Packaged House designed by Walter Gropius and Konrad Wachsmann, and Meudon Housing designed by Jean Prouve. Both examples provided a system with interchangeable components that can be customized by other architects or owners for multiple configurations. However, lacking a well-communicated design interface among end-users, architects, and manufacturers created obstacles for making their dream of factory-made house to be realized.

Today, we are immersed in the digital age that created opportunities never before available to connect information, people, products, and tools in a comprehensive manner. Many industries adopted mass customization concept as their business goal and utilized web media as the communication interface to satisfy their individual client's need. However, the impact of this new paradigm seems to take a longer time to reach architecture, which often lags behind other fields. There are barely a few implementations applying the concept of consumer-driven customization in the housing industry which is more directly related to personal life style. Figure 2.31 shows different degrees of customization approaches in terms of consumer-driven design implementation. This research intends to explore the lack-of-defined area in the early stage of programming and schematic design phase.

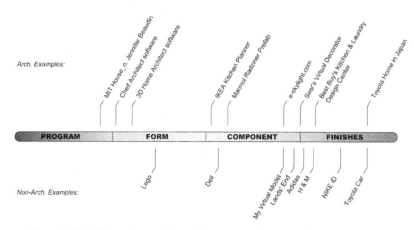

Figure 2.31. Different Approaches of Consumer-Driven Tools for Customization

CHAPTER 3

PROBLEM STATEMENT

If there are so many advantages of prefabricated housing, then why has it not been implemented more today? Besides zoning restriction from most cities in the United States, the "one size fits all" standardized mass production housing statement can not fit today's society and respond to differing local conditions. Although there are more and more emerging architects interested in innovative prefab dwelling, most of them are focusing on design issue with adding the style of modernism, and only a few of their projects provided the customizable options following by the end-user's needs. The current status of prefabricated housing is only being achieved to move the conventional stick-built construction to the factory environment.

3.1   Failure Factors of Prefabricated Housing Industry

Despite the advantages of prefabricated housing introduced in Chapter 1, there are so many factors impeding the industrialization of the building industry. From the contractor's point of view, the economic benefits of industrialization are often marginal. The motive for introducing factory-made housing is not mainly because of economic reason; it is much more likely to be the shortage of experienced workers on the building site and a decreasing willingness to do dirty work (Thillart, 2004). If the building industry ever succeeds in attracting sufficient cheap labor, traditional building methods will undoubtedly play a greater role. Unlike most other industrial products, buildings are static and so tied to the conditions prevailing in the locations where they are "sold". Obviously the building industry can not be transferred to low wage countries, as has happened with other industries requiring extensive manual labor, such as shipbuilding and textiles. Even if it is possible to produce some building components from the factories of the low wage countries, the technical constraints

55

and substantial transportation costs of bring complete three-dimensional building elements to the building site will limit the economic feasibility.

Volume is another key factor to all factory-built house creators, because their business can not survive if the initial cost of the factory can not be paid off by the final housing production output. In the historical review of prefabricated housing, we can see there are many pioneers with great design solutions that failed in their business achievements. Cultural issues also play an important role in the failure factors of prefabricated housing. Normally, people have a poor image of prefabricated housing, and usually recall the image of mobile homes or temporary shelters. The confusion between different terminologies within "prefab family" has been defined in Chapter 2.1. Another observation coming from American culture is that most people prefer to DIY their house work, either new construction or home improvement. If the prefabricated house system is not flexible enough or the components of house are not exchangeable easily, it will be difficult to customize it based on the user's need. Overall, the biggest obstacles of making improvements in this industry are lack of cooperation and remaining outdated technology.

3.2   Limitation of Existing Approaches

The limitation of existing approaches for integrating mass customization concept into modular house design can be reviewed with the following three aspects:

- Specific vs. Holistic: The existing approaches of mass customization implementation in housing industry are more dealing with specific components of prefabricated modular house (i.e. skylight design in Chapter 2.4.2). Without a holistic awareness of building design to construction processes, clients cannot have a big picture of designing their dream houses.

- Product-oriented vs. Client-oriented: Only a few prefabricated housing providers adopted web technology as a communication interface between clients and designers, and the implementation they have achieved is only like a web-based catalog without any intellectual architectural programming process. Most existing prefabricated housing providers address their product's features rather than the client's needs in their web-based design interface.

- No rule vs. Too-much rule of the user's interface: Some consumer-driven home design applications have been introduced in Chapter 2.7 that there are no restrictive rules between each building component when users draw the house plan. Normal people without professional training cannot achieve a logical and constructible design scheme. On the other hand, a catalog-like design interface represents too-much rule in terms of mass customization concept. The existing user's interface did not provide a balance of freedom and restriction in the decision making of design process.

3.3   Scope of Research

The failure factors mentioned above are related with social, cultural, and economic issues. It reflects the fragmentation of overall building industry today. With a personal effort and an academic research output, it is more possible to reveal problematic issues rather than change the entire industry. Figure 3.1 shows the three components: modular system, home design, and digital design application as research study areas. How to integrate these three areas together to become an optimum solution is the goal of research. Although prefabrication is the motive and big concept of this research, prefab technology is such a broad topic with a very sophisticated content. A comparatively narrow area of modular system applied to single family house type has been carefully chosen to be analyzed and reconfigured in case studies. Architectural programming of home design becomes the core of this thesis content and the research must determine the relationship of home design components and modular system configurations. Digital design technology[12] is the tool and method of approach to achieve the goal of mass customization in prefabricated housing design. Hence, this is a cross disciplined research among these three components.

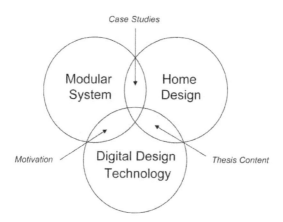

Figure 3.1. Scope of Research and Integration of Multi-disciplines

3.4   Definition of Problems

Through background research, there are a few problems of implementation of mass customization in prefabricated housing that have been defined which must be resolved along with thesis research:

1. Factory-made housing construction method brings efficiency and affordability by mass produced modules. However, prefabricated housing industries failed to address the issues of variability and individual needs from the consumers.

---

[12] Digital design technology is general term and may include computer-aided design, web-based design collaboration and cyberspace, form generation and fabrication, ubiquitous computing, etc.

2. Adding mass customization value into prefabricated housing is a complex undertaking with involving a number of independent parties, like component manufacturers and product providers, etc. There is no guideline for clients to get an overall idea of the project delivery process.

3. Face-to-face meeting time between architect and client is always limited and time consuming for residential projects. How to construct a more accessible consulting tool for customizing factory-built mass produced housing?

4. Most prefabricated housing vendors only achieve in a single design model or very limited design alternative, and rarely address any issues related with client's requirement in a systematic method.

3.5   Objectives of Research

The main goal of this research is to investigate the possibilities of customizing mass housing by internet and prefabrication technology. Figure 3.2 shows the new model of prefabricated housing delivery system and the focus area of this research.

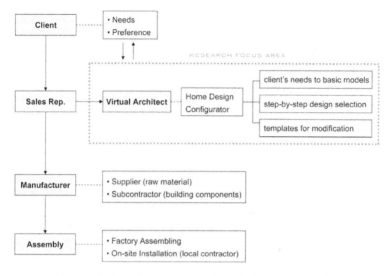

Figure 3.2. Research Focus Area for the Model of Prefab Housing Delivery System

The specific research objectives are stated as follows:

1. To research how to collect and interpret client's need to formulate design options via internet in order to address the issues of individual needs from the consumers.

2. To explore possible combinations of prefabricated modular house according to client's preference.

3. To suggest a framework of mass-customized prefabricated housing delivery system and simulate the process from the definition of client's preference to the ready-to-build housing components.

4. To propose a method of web-based collaboration for the client, architect, and manufacturer to integrate the process of mass-customized prefabricated house online.

3.6   Importance of Research

The importance of this research is that no prior developments have been made that suggest utilizing dynamic questionnaires by web interface to customize modular housing design and select appropriate building components for manufacturing. It tries to identify the client's needs and housing requirements to match the existing prefabricated housing systems plus design modification. The web-based design tool allows clients to see the result of design feedbacks quickly and accessible everywhere.

Besides that, the research tries to explore the possibilities of a new prefabricated modular design system to address client-responsive issues. This approach will suggest a new way to integrate the involvement of end-users into housing delivery process with today's technology.

CHAPTER 4

METHODOLOGY AND ANALYSIS

4.1   Conceptual Framework

In order to achieve the goal of mass customizing prefabricated modular housing, the
prototype model combines the results of two important parts: data collection of client's
requirements and a prefabricated modular system of design configurations. A web-based
prototype can simulate this interaction between clients and the adoptable systems. The
evaluation phase can include a series of case studies to demonstrate and revise the data-input
method within the design interface. Finally, the resultant design can generate building
specifications prepared for the manufacturing and assembly of the products (Figure 4.1).
Warszawski addressed five tasks for computer-aided design and planning in prefabrication
plant: (1) Input of architectural design, (2) Pre-estimating, (3) Outline of the elements, (4)
Detailed design, and (5) Production planning (Warszawski, 1999). This research will be
focused on the input methods of the end-users rather than relying on architects for finding
suitable design solutions in the prefabricated housing.

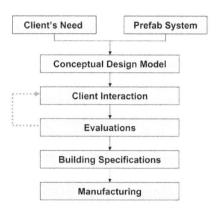

Figure 4.1. Conceptual Framework of the proposed i_Prefab Home System

4.2   Proposed Model

Presently, only five percent of people in the United States typically hire an architect and pay them to design and build a home which is tailored to their preference[13]. Instead, many people purchased stock house plans and hire a home builder for their new single-family houses. Besides the architect's fee, clients also need to wait an interminable time for the completed process of design and construction. Factory-produced prefabricated housing systems have previously tried to solve this problem. However, most systems failed to address the issues of variability and individual needs. Plants closed because they could not achieve the large enough market needed to reduce the costs, and the prefabricated housing provided less flexibility and could not compete with the stick-built housing market.

Now, advanced digital technology makes it possible to communicate design ideas and concepts to others more effectively. The project delivery process leads itself to customization, embodying principles of lean production (Pine, 1993), flexible computer-integrated design interaction with clients, and reduced cycle times; all effecting rapid response between consumers and producers (Figure 4.2). Demand-to-order is not a dream for prefabricated housing industry anymore. As long as people are motivated to accept this new concept, prefabricated housing will shift from the stereotype of "factory-like" repetitive industrialized products to flexible and customizable humanized products.

----

[13] 2003 AIA Firm Survey, http://www.aia.org/press_facts

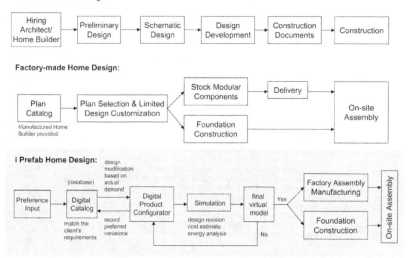

Figure 4.2. Existing Models and Proposed Model

4.3   Dwelling Type Consideration

There are many types of dwelling. Regardless of the different cultures and styles in the world, the common types by units are single-detached, semi-detached, duplex, row house, 3-store walkup apartment, mid-rise and high-rise condominium. This section explains the reason of the selected dwelling type and defines the scope for the further analysis and case study.

America is a single-family-house country, and there are more than two-third of people live in this type of dwelling. Besides the privacy and independence of the yard, the building itself has the freedom to express the character of design without any shared wall issue with the neighbors. Figure 4.3 indicates the single family detached housing type is about 63% of overall housing types. If the single-family house data include single family attached (5%) and mobile home/trailer (7%), this type of dwelling is about 75% of existing housing types and represents the most popular housing type in this country. Although the case study of this

64

research is using the most popular housing type in the U.S. with a simplified condition, the results of these design experiments and analysis can be applied as a universal idea. Figure 4.4 is a photo to show the most common different housing types in the world.

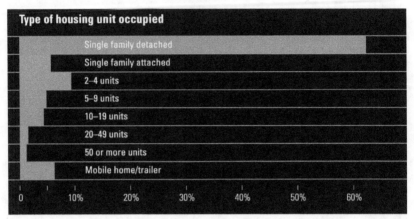

Figure 4.3. Type of Housing Unit Occupied in the U.S.[14]

---

[14] U.S. Department of Commerce, Economics and Statistics Administration, Bureau of the Census; U.S. Department of Housing and Urban Development, Office of Policy Development and Research. American Housing Survey for the United States in 1995. Current Housing Reports H150/95RV. Washington DC.

Figure 4.4. Different Types of Housing

In this research, the hypothesis design model for case study is single family detached house with two different site conditions: urban in-fill and suburban location. All case studies apply the prefabricated technology and modular system with the limitation of lot size in both urban and suburban conditions. Similar to bungalow, the case study house for the urban condition is a one-story house with a small square footage, and design to provide affordable, modern housing for young couples as their first house. In order to simplify the result of system variations and comparison of urban and suburban restrictions, the scope of analysis is designed to focus on two-bedroom single-family house as urban site case study, and four-bedroom single-family house as suburban site case study. The detailed scope of analysis and prototype design has been defined on Chapter 4.4. The web-based prototype system is to proof the methodology of mass customizing modular house with internet technology and questionnaire approach, not aims to create as commercial software.

66

4.4   Defining the Scope of Analysis and Prototype Design

- Dwelling Type: Single-family house

    - Urban in-fill detached house (Standard Lot 25'x125')

    - Suburban detached house (Lot 100'x125')

- Type of Module:

    - Fully Modular (analysis & case study)

    - Sectional (analysis only)

    - Component (analysis only)

- Market: Middle-income House

    - Cost: ~ $ 200,000 (price not including the land)

- Number of Rooms:

    - 2 bedrooms + 1 bath

    - 4 bedrooms + 2 baths

- Parking/ Garage: for suburban detached house only

- Number of Floors: 1 + basement

- Square Footage/ Unit: 1000 ~ 2500 square feet

- Method of Transportation: Truck & Train

- Site Location: No specific site; assume with a prepared foundation

- Potential Usages of this Prototype:

    - Affordable single-family house

    - For the housing needs after disaster (not as temporary shelters)

    - A new prototype for developing countries

4.5   System Flow of Proposed Model

The internet-based prefabricated modular house design advisory system (i_Prefab) has two modes, basic and professional. The basic interface allows customers to query the suitable design options and customize the detail components as needed. The professional interface is designed for architects and home builders with the capability to organize the building site issues and review customer's final selections. It also allows advanced users to exchange some digital data through different digital design applications in order to collaborate with engineering consultants and manufacturers. Figure 4.5 shows the system flow chart of proposed model.

The basic mode of the proposed system starts with a dynamic questionnaire – a series of questions to address the customer's household profile, lifestyle, basic site context information, space requirements, and design style of home construction. Then, the advisory system will incorporate the customer's input to find the most suitable design options from its database. Once the customer selects the matched model from suggested design options, the customization process begins. The online configuration tool provides the functionality to change or add design elements based on user's preference. It contains four categories: space planning, construction, appearance, and appliances/equipments. Space planning is the most important component of these categories. Each individual room is treated as a replaceable "block", and can be exchanged by a different layout of the same space type or a trade-off with different usage. There are some restrictions assigned to these room blocks to control the associated wall types which have to be compatible with their neighbors when those rooms have been rearranged. Room blocks can also be treated as additional space to extend the boundary of a suggested design floor plan.

Construction is the second category of the configuration tool. This is a general term for laypeople, and basically indicates any major change after the design plan has been decided. Regularly, the foundation of a prefabricated modular house will be prepared by the local contractor on its building site. However, it is very important to understand the type of foundation (i.e. pier, strip or basement) to make space planning adjustment.

Appearance includes surface material, color and texture selection of exterior and interior components. Most of existing home catalogs and sales center have already achieved this service. The last category, appliance, may apply from kitchen appliances (refrigerator, dishwasher, oven and vent) to laundry and air conditioning equipment. Recently, the concept of sustainable design is getting more and more popular. Integrating solar panels and wind turbines with building design is no longer limited to commercial buildings. Energy renewal is becoming a desirable solution for residential projects and it can be applied for prefabricated modular housing as a new feature of the energy appliance option.

At the end of customization process, the advisory system will provide a price quote for review. Once the customer is satisfied with the design and estimated cost, he or she can meet with architect consultant to schedule a showroom visit for further modification. On the other hand, the professional mode of the proposed i_Prefab system is more straight forwarded from site planning to system and component selections. The important phase is when customer-driven design model is merged to the professional mode for architect's review. All of the proposed modular house components are made by Building Information Modeling (BIM) application as a virtual building data file for professional analysis and simulation before factory manufacturing.

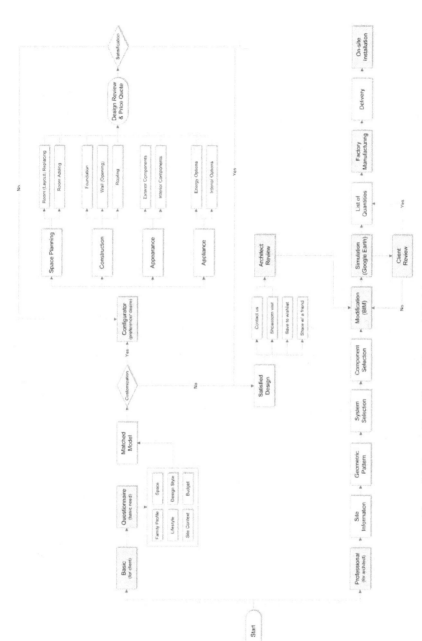

Figure 4.5. System Flow Chart of Proposed Model

70

4.6   Analysis of Existing Modular House Systems

In order to create a reasonable consumer-driven design advisory system for prefabricated modular housing, a fully understanding of the modular systems is the first step. In the background review, four different types of prefabricated housing systems have been introduced as fully modular, sectional, component, and hybrid. Since the hybrid system is a combination of the other three basic modular systems, it has been eliminated in this case study.

The reason to select prefabricated housing systems from the existing vendors is to understand the available systems in the market. Since there are many decent design products already have been proposed by architects and fabricators, this method can focus on the issues of consumer-driven interface and design customizing process rather than invent a new modular system. Furthermore, these selected prefabricated design products have been pre-engineered in the factory and followed the building codes as realistic elements. The selected products can also be used as examples in the interface prototyping of this research. The result of analysis will create different configurations and can be applied for the future development of modular systems that based on the consumer's need or adaptability of the modular components.

4.6.1 Criteria of Selection. Generally speaking, the existing prefabricated modular housing market can be divided into two categories based on the appearance: traditional and modern. Traditional modular houses have been commercially available for decades, and most of projects are indistinguishable from traditional site-built houses. Traditional style of modular housing provides a various degree of customizable options to simulate as site-built construction. Unlike decorative traditional style, the current movement of prefabricated housing adds Modernist aesthetics to make a "volumetric" approach, which is more emphasis on spatial composition and the relationship with its site context.

Fabprefab.com, established in 2003 and created by Michael Sylvester is a web resource dedicated to tracking developments in the market for modernist prefabricated dwellings (Figure 4.6). He saw commercial possibilities in the suddenly high-profile modular-construction industry. His web showroom exhibits designers on the cutting edge of prefabricated and modular construction, but only those who are doing it in the language of Modernism. There is an increasing number of architects are discovering the unserved niche in the residential market – modern houses for consumers who cannot afford the one-off, expensive architectural masterpiece (Hart, 2003).

There is a category called "fablist" from fabprefab.com. All projects in this category have to meet three criteria: (1) have had a prototype or finished example of the dwelling constructed; (2) are actively marketed as a product/solution by a vendor; (3) are available for purchase in North America. In this research, four design products (Table 4.1) have been selected for analysis, and those selected products have highly configurable possibilities and cover three different modular types (fully modular, sectional, and component) that have been introduced in Chapter 2.3.

Instead of one design product/ system per one modular type, there are two projects have been selected in fully modular category for analysis. The reason is the lack of one-level

72

example for 3-bedroom and 4-bedroom in current weeHouse modular system. In order to avoid the user's confusion of interface prototyping, focusing on one-level configuration has been set as research perimeter.

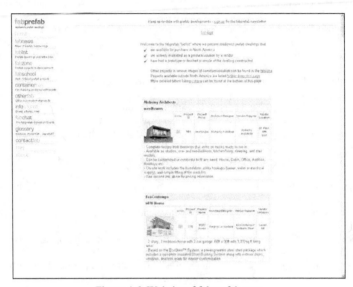

Figure 4.6. Website of fabprefab.com

Table 4.1. Selected Modular Design Products and Systems for Analysis

Modular Types	Project Name	Architect/ Designer	Vendor/ Supplier	Location of Architect	Basic Module
	weeHouse	Alchemy Architects	Alchemy Architects	St. Paul, MN	14' width
Fully Modular	The Modern Modular Homes	Resolution:4 Architecture	Empyrean International LLC, Acton, MA	New York, NY	16' width
Sectional	kitHAUS	Sandonato & Wehmann	kitHAUS	Los Angeles, CA	16' width
Component	FlatPak House	Charlie Lazor/ Lazor Office	Empyrean International LLC, Acton, MA	Minneapolis, MN	8' panel

4.6.2   weeHouse.   The weeHouse prefab system is based on a modern aesthetic, efficient use of space, and intelligent adaptation of building technology. From small retreats like the original weeHouse (Figure 4.7) in Wisconsin, and the Marfa weeHouse to larger residential and commercial weeHouses currently in progress around the country, the system is adaptable to a wide range of needs[15]. The weeHouses arrive on site ready to live in. The only things left for the contractor are the foundation, fitting & seaming of the modules, utility hookups, and in some cases cabinetry.

Figure 4.8 illustrates the system variation of weeHouses as three groups: (1) weeHouse Solitaires; (2) weeHouse Companions; (3) Not-so-weeHouse. The modules of Solitaires are single-unit modules, and one modular unit can be sustained as one living unit (Figure 4.9). Companions play as parts of living units or special use like a modular stair. Any module from weeHouse Companions has to be connected with other modules to be a complete living unit. The last one, Not-so-weeHouse, is a combination of everything mentioned above, and up to 2-story construction and 4-bedroom configuration.

Figure 4.7. The First Prototype of weeHouse

---

[15] weeHouse Brochure, http://weehouses.com/new%20pdfs/weeHouse%20Brochure%209-06.pdf

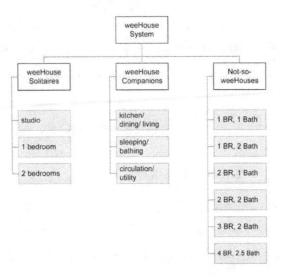

Figure 4.8. The System Variation of weeHouses

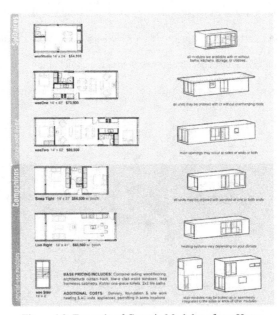

Figure 4.9. Example of Generic Modules of weeHouse

4.6.3   The Modern Modular Homes.   With a modern modular design methodology, New York based design firm, Resolution: 4 Architecture won Innovation in Housing Design Award from the American Institute of Architects (AIA) in 2006. Taking full advantage of the flexibility of their systemic approach, these houses are based on a set of bar-like modules that can be joined to make H, I, L, T, and Z configurations with at least 35 variations in terms of length, depth, and height (Figure 4.10). The firm's competition-winning Dwell House (Figure 4.11), completed in 2004 for Dwell magazine, set the stage for many other modular homes[16].

Figure 4.12 divides the system variation of the Modern Modular Homes into three categories: (1) by floor plan shape; (2) by space zone; (3) by room number. Similar to weeCompanions, architects from Resolution: 4 Architecture divide the space zone into three different types of essential module and result in their house plans that expressed as an arrangement of communal modules (living and dining spaces and kitchens), private modules (bedrooms and bathrooms), and accessory units (storage and decks). By bringing these elements together in ways that reflect the client's unique needs and site conditions, they are able to leverage the benefits of factory-built construction with the advantages of a custom-designed residence.

---

[16] Resolution: 4 Architecture, official website http://www.re4a.com

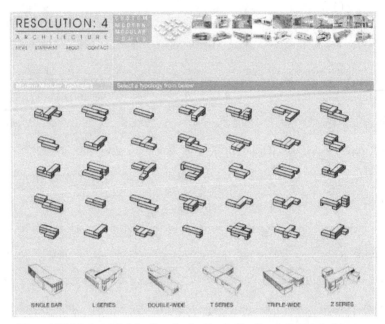

Figure 4.10. Modern Modular Typologies Created by Resolution: 4 Architecture

Figure 4.11. The Dwell Home Designed by Resolution: 4 Architecture

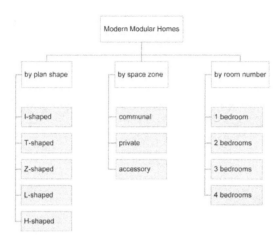

Figure 4.12. The System Variation of the Modern Modular Homes

4.6.4 kitHAUS.   Kithaus is a perfect modular house that can be used as a studio of vacation. The best thing about Kithaus is that users have the freedom to design the space any way they want because the modules are built with a patented MHS (Material Handling System) frame with sandwiched insulated panel exterior, clad in wood. All components pre-cut, marked and drilled and can be assembled in days. Many configurations are available using assorted modules (Figure 4.13, 4.17).

Although this is the only system that achieved sectional spatial arrangement from fablist category of fabprefab.com, it is still very controversial to identify Kithaus as "sectional" modular type. Since there is no real "pure sectional" modular house has been built, the analysis result will not transfer for prototype use. However, sectional modular design approach is very attractive, and has been expressed through many design competition works (Figure 4.15).

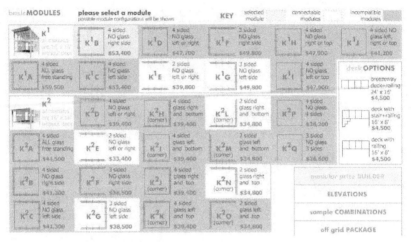

Figure 4.13. Matrix of The Basic Modules of kitHAUS

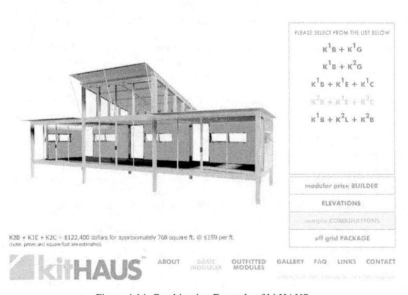

Figure 4.14. Combination Example of kitHAUS

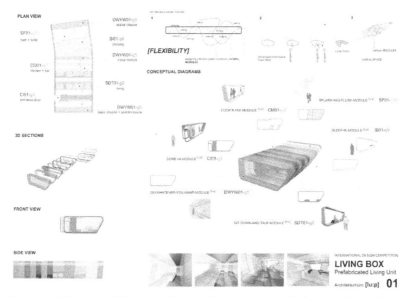

Figure 4.15. Conceptual Example of Sectional Approach from Living Box Competition

4.6.5   FlatPak House.   Charlie Lazor, principal of Minneapolis-based Lazor Office, began

his exploration of prefabrication in 2002 through the creation of a home for his family. The

resulting prototype as a two-story, three-bedroom, three-bath house with a separate study and

guest room, was completed in 2004 and launched the FlatPak series. As the name suggests,

the system evokes a do-it-yourself attitude by offering owners a wide range of choices and a

hand in the layout of their spaces. For instance, one could choose glass, wood, concrete, or

metal panels to create a wall (Figure 4.16, depending on function and location. In this

scheme, the homeowner does not assemble the house but rather becomes an active participant

in its design. Numerous configurations are possible because FlatPak is based on a simple 8-

foot-wide, 1-story-high wall panel. There is no fixed length and up to four stories are

possible. The prototype of FlatPak house (Figure 4.17is produced by Empyrean, an

experienced company that specializes in the construction of modern, prefabricated houses.

Costs average $175–$250 per square foot and reflect a complete house package that also includes design, engineering, and project management services[17].

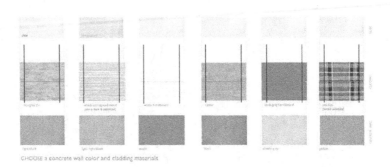

Figure 4.16. Panelized System in FlatPak House

Figure 4.17. The First Prototype of FlatPak House

4.6.6   Space Configuration Analysis.   The purpose of this analysis is to test the possibility of generating more spatial configurations beyond the current products provided by modular housing vendors. Despites of many architects and manufacturers currently approach this direction of prefabricated housing, not many house design products provide more than

[17] http://www.thedwellhomesbyempyrean.com/Preview3/Designers/Flatpak/03_flatpak.html

one variation type or enough flexibility for consumer to make further customization rather than selecting finishes based on the fixed building geometry.

Standard Bar of Modern Modular Homes designed by Resolution: 4 Architecture is the simplest 2-bedroom modular house system to be tested for the variation of spatial layout (Figure 4.18). The design configuration of modular system starts from a 16' x 60' two-bedroom unit. This design provides the most economical and efficient solution for a modern modular prefabricated home. Within a compact single unit, the home contains two well-sized bedrooms with ample storage space and can be carried by one truck to the building site.

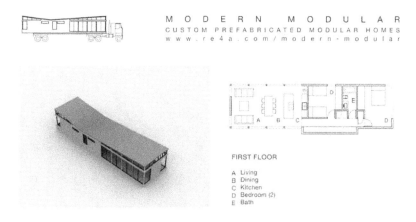

Figure 4.18. The Original 2-Bedroom Design Plan from Modern Modular Homes

Figure 4.19 shows the tree diagram of 2-bedroom case study modular groups and modular units. The two-bedroom modular house has been divided into "public space" modular group and "private space" modular group. Then, both modular groups can be sub-divided into six modular units. In Figure 4.20, the A1 floor plan shows the public space (c1) on the left and the private space (p1) on the right. After the modification, the client has two options to determine the first space from the front door is living room or kitchen. Moreover, there are two variations in the private space: bathroom locates in-between two bedrooms or two bedrooms link together.

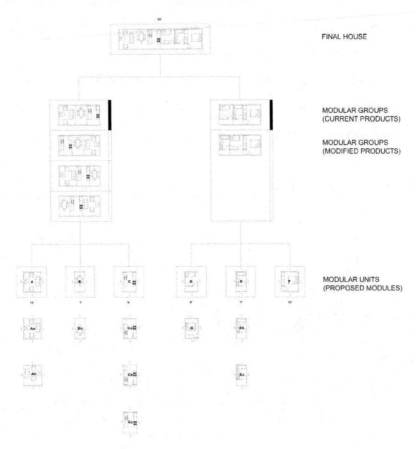

Figure 4.19. Diagram of 2-bedroom Modular Groups and Modular Units

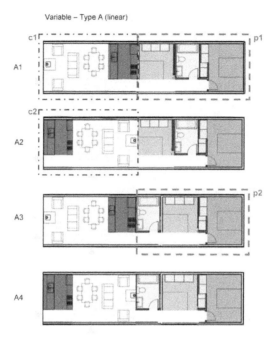

Variable – Type A (linear)

Figure 4.20. Space Reconfiguration Study for Narrow Lot

The second step is to test the space configuration within square geometry. Unlike linear configuration, square floor plan can reduce maximal interior circulation in the plan B1 (Figure 4.21). The plans B2 and B3 are special cases of non-linear configurations of the public space and private space. After interlocking of these two spaces (c3 and p3), there is a potential room for home office or guest room in the plan B2 and B3 which is close to the public space. B3 plan is a semi-revised plan of B2 to provide more privacy for the master bedroom.

Figure 4.21. Space Reconfiguration Study for Wide Lot

After the basic spatial composition exercise, the whole house can be sub-divided by its real space units, like living room, dining room, kitchen, bathroom, and bedroom. There are four different reasonable configurations in the public space of 2-bedroom modular house as below:

1. Living – Dining – Kitchen

2. Kitchen – Dining – Kitchen

3. Living – Kitchen – Dining

4. Dining – Kitchen – Living

Plus, there are two conditions of the private space:

1. Bedroom – Bathroom – Bedroom

2. Bathroom – Bedroom – Bedroom

Therefore, the system variation may end up with eight different compositions following the same linear building geometry. Figure 4.22 is the matrix of space adjacency of these six modular units.

Next step is to label the identification tag for each modular unit with letters A to F, and numbers 1 to 4 on four sides of each unit. After carefully rearrange the basic module with architectural considerations and minor modifications of some modular units, like mirrored unit or adjusted adjacent wall conditions, the original six basic modular units can generate as many as 22 different configuration types of a 2-bedroom, one-story single family house for the different people and site variations (Figure 4.23 ~ Figure 4.27).

ZONE			PUBLIC			PRIVATE		
			LIVING	DINING	KITCHEN	BEDROOM	BATH	M. BEDROOM
	ROOM	ROOM						
PUBLIC	LIVING		NO	YES	YES	YES	YES	NO
	DINING			YES	YES	YES	YES	YES
	KITCHEN				NO*	YES	YES	YES
PRIVATE	BEDROOM					YES	YES	YES
	BATH						YES	YES
	M. BEDROOM							NO

NOTE: UNLESS IT WILL BE THE OPTION OF ENLARGED SPACE (X2), AND INGORE THE LAYOUT OF KITCHEN

Figure 4.22. Matrix of Modular Units

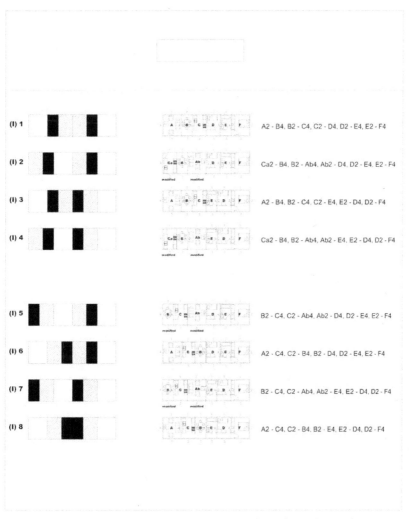

Figure 4.23. New System Variations based on the Basic Modular Units

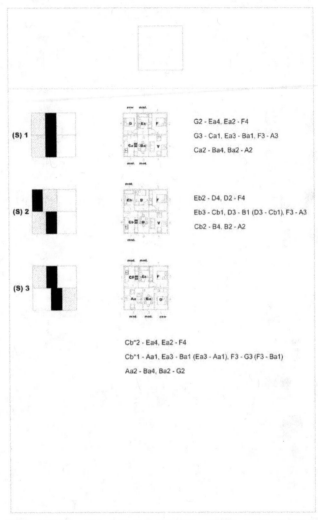

Figure 4.24. New System Variations based on the Basic Modular Units

89

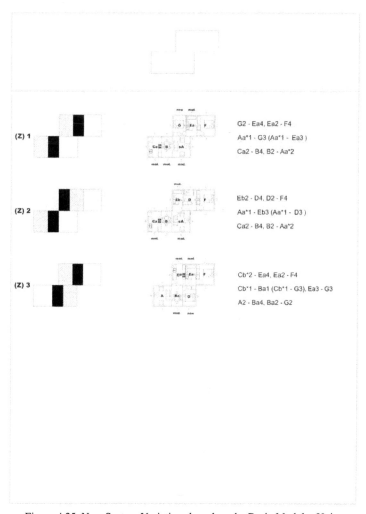

Figure 4.25. New System Variations based on the Basic Modular Units

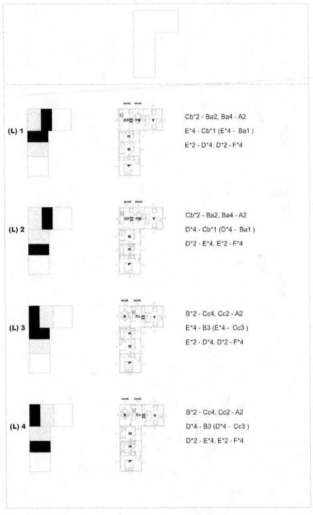

Figure 4.26. New System Variations based on the Basic Modular Units

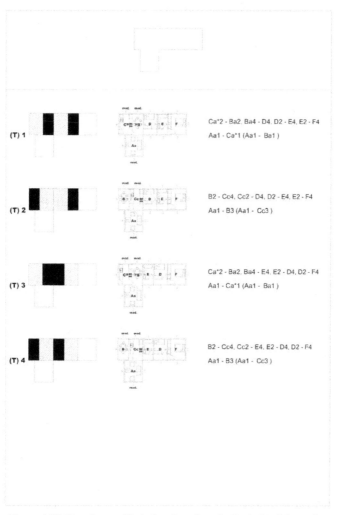

Figure 4.27. New System Variations based on the Basic Modular Units

After 2-bedroom modular house configuration analysis, the research moves to 4-bedroom modular house to examine the bigger lot condition in suburban. The reason to skip 3-bedroom study is because the case is just in between 2-bedroom and 4-bedroom, and the

92

study result can be applied for this transitional case. The house with four bedrooms is more typical in suburban and acceptable for most family size with a couple, two children, and another flexible room as guest room or home office. Unlike 2-bedroom design experiment exercise, the 4-bedroom study and analysis are focused on standardization of modular components.

The Modern Modular Homes designed by New York based design firm, Resolution: 4 Architecture has been introduced in Section 4.7.3. They won AIA Innovation in Housing Design Award due to the flexibility of their systemic approach and variety of modular house design configurations. This analysis examines 5 different models of 4-bedroom, 1-story Modern Modular Homes. How to minimize the sharing standard modules from the existing variable design configuration is the first task.

Refer to Pine's theory (1993) discussed in Section 2.4.1, component-swapping modularity is the ability to switch options on a standard product. Modules are selected from a list of options to be added to a base product. Figure 4.28 and Figure 4.29 show two examples of component-swapping mass customization method for group module and individual space module. The floor plan variation can be determined from the combination of the essential modules like communal modules (living, dining and kitchens), private modules (bedrooms and bathrooms), and accessory units (storage and decks). In order to achieve the detailed level of customization, each individual room should be independent as a room module to act as a component-swapping element.

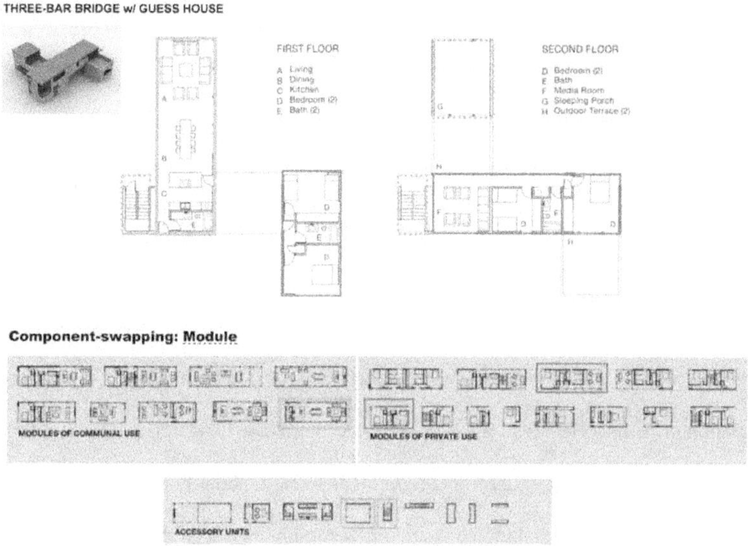

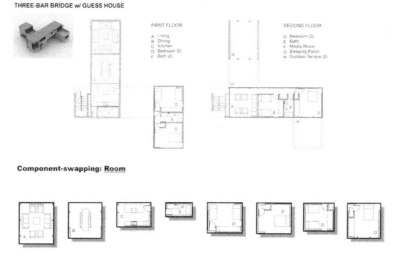

THREE-BAR BRIDGE w/ GUESS HOUSE

FIRST FLOOR

A Living
B Dining
C Kitchen
D Bedroom (2)
E Bath (2)

SECOND FLOOR

D Bedroom (2)
E Bath
F Media Room
G Sleeping Porch
H Outdoor Terrace (2)

Component-swapping: Module

MODULES OF COMMUNAL USE

MODULES OF PRIVATE USE

ACCESSORY UNITS

Figure 4.28. Component-swapping Mass Customization Method for Group Module

THREE-BAR BRIDGE w/ GUESS HOUSE

FIRST FLOOR

A Living
B Dining
C Kitchen
D Bedroom (2)
E Bath (2)

SECOND FLOOR

D Bedroom (2)
E Bath
F Media Room
G Sleeping Porch
H Outdoor Terrace (2)

Component-swapping: Room

Figure 4.29. Component-swapping Mass Customization Method for Room Module

Figure 4.30 shows the comparison of basic information and special features in 4-bedroom, 1-story Modern Modular Homes. These models are sort by square footage from

94

1895 S.F. to 2282 S.F., and most of them have a clear division of public and private spaces. Besides the view factor of the site context and building geometry, the most common feature is suitable for a large family, or work as a shared duplex residence.

After analyzed the space configuration of these five models, Figure 4.31 illustrates the standard products of Living Modular Group (C3, C4, and C8) and Sleeping Modular Group (P1, P6a, P6b) in the upper-left corner, and system variations applied directly from the vendor and after modification. The result tells us that 4-bedroom in suburban condition has freedom to play the plan geometry since there is less restriction of the site if comparing with the urban narrow lot. However, it is important to control the standardization factor to reduce the cost and time spending of customizing prefabricated modular houses. The prefabricated housing design is not acceptable in a single product with only one rule, but it cannot achieve the efficiency by too many possibilities with no rule during the mass customization process. The optimized solution is to keep a minimal number of standardized group modules which are capable of generating a maximal number of reasonable design schemes.

Compare Selected Models (based on **4 bedrooms. 1 story**)

model photo	basic information	special features
	**THREE-BAR 'T'** (price start from $$) Square Footage: 1,692 Bathroom: 2 Living Area: 1 Dining Area: 1 Exterior Dimensions: 45'X78'	• Central living space has **180° view** to the landscape and perfectly matched for use on a site with a beach or large acreage • Suitable for a large family, or work as a shared duplex residence
	**2 BAR SLIP** (price start from $$) Square Footage: 1,895 Bathroom: 2 Living Area: 2 Dining Area: 1 Exterior Dimensions: 32'X94'	• Two living units are anchored by a central living space with windowed **views in every direction** • Suitable for a large family, or work as a shared duplex residence
	**3 BAR SADDLEBAG** (price start from $$) Square Footage: 1,909 Bathroom: 2 Living Area: 1 Dining Area: 1 Exterior Dimensions: 48'X60'	• Separate **public and private spaces** in a compressed volume that works to create a lateral spaciousness.
	**3 BAR PINWHEEL** (price start from $$$) Square Footage: 2,072 Bathroom: 2 Living Area: 2 Dining Area: 1 Exterior Dimensions: 55'X70'	• **Courtyard** created by the building shape to form an individual exterior space
	**3 BAR DUPLEX** (price start from $$$) Square Footage: 2,282 Bathroom: 2 Living Area: 1 Dining Area: 1 Exterior Dimensions: 45'X60'	• A clear division of **public and private spaces** • A central service unit separates the living and bedroom modules • Suitable for a large family, or work as a shared duplex residence

Figure 4.30. Comparison of 4-bedroom Series of Modern Modular Homes

96

Figure 4.31. Analysis of System Variations, 4 Bedrooms

4.7   Decision Tree

Since the proposed model of the advisory system needs to provide the option of consumer driven participative design, a prototype interface which can guide consumers to make a clear decision has to be constructed. A decision tree is an idea generation tool that is used to identify the strategy most likely to reach a goal. It can also be treated as a support tool of data mining from the knowledge database, because the users can achieve the target item by passing a series of decision-making nodes. Personal Brain, created by TheBrain

Technologies, is an easy-to-use system for organizing and sharing information. The first prototype adopts this application as a platform for a working model. The interface is divided into two portions: decision tree navigation and an input/ output dialogue window. The input/ output dialogue window includes the questionnaire (Figure 4.32) and design suggestions (Figure 4.33), which has been introduced in Chapter 4, Section 4.6. By utilizing the decision tree feature of its interface, the questionnaire answering process will be represented as decision trees in the upper views of the interface for guiding users to find the suggested design solution. The suggested design solutions would reflect the available systems of a specific vendor within the existing market.

A traditional decision tree pattern even shows the decisions which seem simple to the customer can create a large number of potential system states for the developer and a similarly large number of different connections and opportunities for failure (Thillart, 2004). For example, a decision tree with two systems states can generate eight systems states after only three decisions provided by the customer (Figure 4.34). However, some system states may repeat and the overlapping relation links will form as a decision network. The optimal goal is to keep a minimal number of different system components which are capable of generating a maximal number of useful product variations.

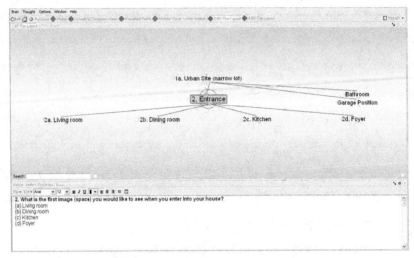

Figure 4.32. Decision Tree Navigation (above) and Questionnaire (below)

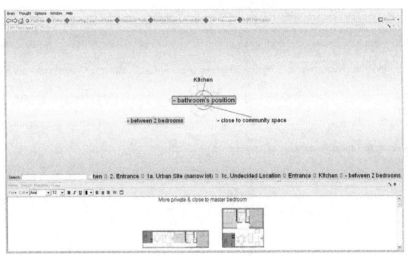

Figure 4.33. Decision Tree Navigation (above) and Design Suggestion (below)

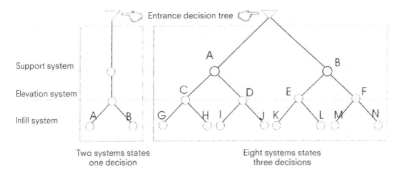

Figure 4.34. Decision Tree of Mass Customization Modeling

The decision tree navigation interface is very intuitive, and customers can understand where they are in the decision-making process and the hierarchical relationship with the other choices. The following case studies are the first step of programming in modular house design. The questionnaire can define how many units (bedrooms) may suitable for the customer as a recommendation. Besides sleeping arrangement of house living members, the decision tree can continually grow up when the basic required units mix with the additional spaces generated by the lifestyle questions. The four different cases of household profile are (1) single, (2) couple with children, (3) single parent with children, and (4) couple without children. Each category can be subdivided into many different conditions. For example, the category of couple with children have following questions like "How many children do you have?" and "Are you planning to have another child within five years?" in order to determine there is a requirement of a spare room for the coming baby. The category of couple without children is more complicated as four different options: (1) older couple who never had children, (2) older couple living with "empty nester", (3) young couple plan not to have children, (4) young couple plan to have children. Since buying a new house need to concern about the future expansion of family members and alternative functions related with personal

100

life, the proposed method is absolutely based on the client's needs rather than market trends. The optimized goal of customization of modular houses is not only for replacing components or selecting finishes, but also provides options to customize space and interpret the client's needs. Figure 4.35, Figure 4.36, Figure 4.37, and Figure 4.38 show four different scenarios of decision-making process with the help of the decision support tool.

Scenario 1: Single

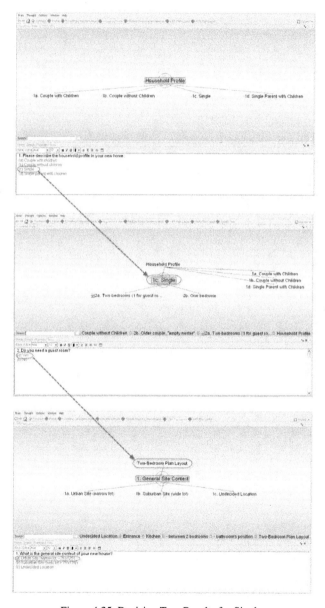

Figure 4.35. Decision Tree Results for Single

Scenario 2: Young couple with one child

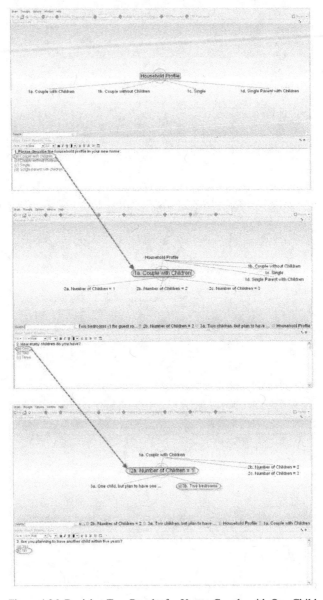

Figure 4.36. Decision Tree Results for Young Couple with One Child

Scenario 3: Single parent with one child

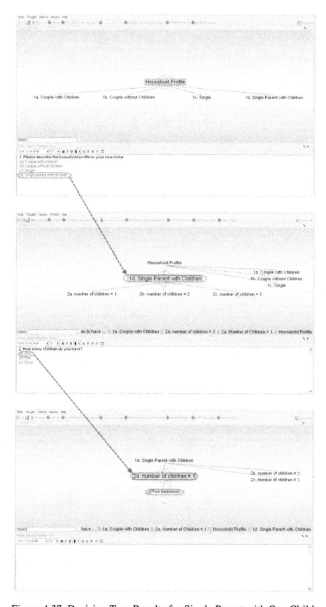

Figure 4.37. Decision Tree Results for Single Parent with One Child

Scenario 4: Retired old couple

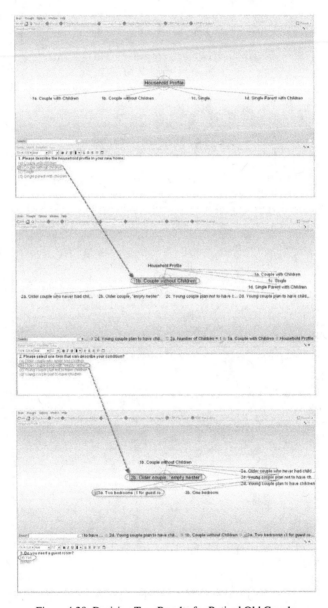

Figure 4.38. Decision Tree Results for Retired Old Couple

4.8   Digital Questionnaire

A questionnaire is an instrument that employs a predetermined set of questions to be answered by a respondent. A survey is the administration of the questionnaire to a group of respondents. It can be administered by an interviewer who records the responses, or given to the respondents to answer at their convenience (Hershberger, 1999). A more common method of conducting housing research is to formulate and organize questions and then distribute questionnaires to residents, because questionnaire studies can suggest non-measurable architectural attributes and describe qualitative values of home design concept to residents (Nylander, 2002).

Mailed-in questionnaires have a characteristically low response rate. Questionnaires are not very good for eliciting new or general ideas or for uncovering issues or doing background investigations because the instrument itself affects the response (Duerk, 1993). The proposed digital questionnaire model links a serial of pre-established answers that define the architectural implementation from its database, and the users will receive a real-time feedback to evaluate home design solutions from the digital interface.

The engagement of internet with pattern book concept can create a power of e-commerce for housing industry, but sometimes the end results of web surfing cannot fit the client's spatial needs. Unlike the other industries (shoes or watch), a suitable house design is not only judged by its appearance or architectural style, but also involves a series of architectural programming phases. Figure 4.39 demonstrates that by reversing the sequence of choosing a product image to get spatial features and functional details, a knowledge-based questionnaire can be a new format to collect the client's input. The interactive web interface will provide advisory design solutions based on the client's need. The main goal of this research is to investigate the possibilities of customizing mass housing by internet and prefabrication technology beyond the finish material selecting process.

106

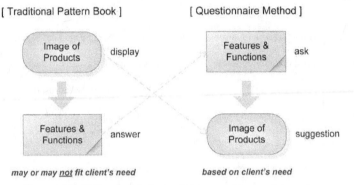

Figure 4.39. Difference Methods of Consumer-Product Interaction

The proposed digital questionnaire model links a series of pre-established answers that define the architectural implementation from its database, and the users will receive real-time feedback to evaluate room layout and home design solutions from the digital interface. From general spatial need to detail preference, there are four different levels of questionnaire to be developed as the programming of this prototype system: (1) Generate a list of required spaces, (2) Determine each room size and relationship of plan by function, (3) Define the detail layout of individual spaces and the development of the plans and elevations, and (4) Customize material and color selections for exterior and interior components, Figure 4.40.

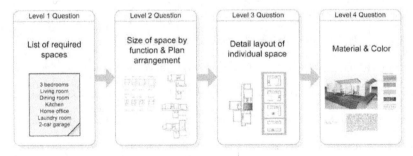

Figure 4.40. Four Different Levels of Questionnaire

The method of questionnaire can be very effective in determining whether or not most respondents share the viewports of those initially interviewed. Like the interview, and unlike observation, the questionnaire can be used to get at why people do what they do, what they think works well or poorly, and how they think something might be done better – but only within the limits of the alternatives actually provided in the questions (Hershberger, 1999). Actually, the limited alternative options provided in the proposed model are represented the available modular systems and components in the existing market. With architect's modification after the chosen model from the advisory system, fully customized design solutions can beyond the limits of digital questionnaire.

The first series of questions are trying to identify the household profile. By answering the household type and how many people in your new home can generate the basic requirement of spaces and sleeping arrangement. Furthermore, the questionnaires of life style, architectural options, site context, and budget can help clients to provide more information and reflect the detail of spatial needs in a short period of time (Figure 4.41). For example, the eating style of the client's family may determine the size, layout, and location of the dining area. If the client needs to work at home quite often, a home office or a den has been considered. All human-machine interactive results can be viewed as design references. The clients have a right to revise any modular component during the design trade-off process. Finally, i_Prefab advisory system will provide available design suggestions from its database. There are two different versions of prototypes documented in Appendix A and B.

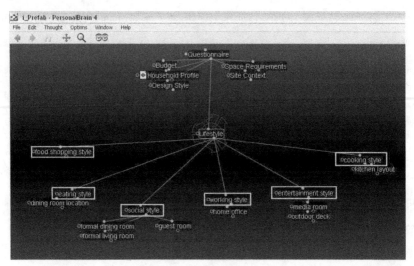

Figure 4.41. Questionnaire of Lifestyle and Corresponded Spatial Issues

4.9   Systematic Approach of Unit, Module, Component and Detail

This section focuses on categorizing different levels of modular house components as a virtual inventory to integrate with the client's input by digital questionnaire and mass customization process of the proposed i_Prefab system. The American architect Stewart Brand, who is specialized in recycling buildings, draws up his own system of categories. He divides buildings into site, structure, skin, services, space plan and stuff (Brand, 1994). Figure 4.42 shows there are five layers of building components besides site element. Here, the "structure" means the load-bearing elements. The "skin" means the covering and protecting membrane, and the "services" means the mechanical, electrical and plumbing system. The "space plan" relates to furniture layout or the space adjacency issue. The "stuff" is an abstract idea and can be interpreted as the space activity which reflects the client's needs.

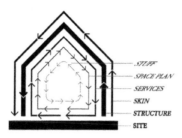

Figure 4.42. Five Layers of Building Components by Architect Stewart Brand

Kent Larson as the MIT House_n research group director simplifies the building components as two main physical components: chassis and infill. As mentioned in the literature review in Chapter 2.5.1, the chassis is a standardized and uniform structural system for the building which allows it to be mass-produced. The infill includes interchangeable components like the floor, wall and ceiling that can be defined by the user based on their preference and needs. Figure 4.43 is a diagram created by Kent Larson's student, Tyson Lawrence to explain the content of the standardized chassis and adaptable infill in the mass customized housing.

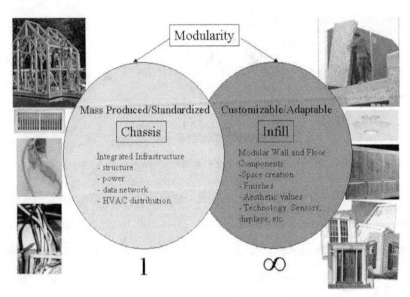

Figure 4.43. Content of Chassis and Infill in MIT House_n Proposed System

The MIT House_n's Chassis and Infill system can be viewed as a modern version of a previous Support and Infill system, which was proposed by John Habraken, chair of the MIT Department of Architecture from 1975-1981. He described support as the communal part of the building containing the structure, services, etc., and infill as the private portion of the building that is tailored to the needs of the occupant and changed over time. Habraken's influential book, "The Systematic Design of Supports" led to what is now called Open Building (Lawrence, 2003) which is popular today in the Netherlands, Japan, and Finland.

The previous two examples focus on the concept of a specific building system and components. However, i_Prefab system is about a housing delivery process of mass customized modular housing driven by the client's input and design suggestions which then determine factory manufacturing and on-site assembling. The research of this systematic approach expands the area of physical building layers to include the abstract value of architectural programming. This modular house design can be divided into four different hierarchically categories: unit, module, component, and detail. The matrix of Figure 4.44 shows the integration of four different levels of the digital questionnaire and four different categories of the modular house system. This matrix of information then becomes the framework of constructing the database for the i_Prefab website interface.

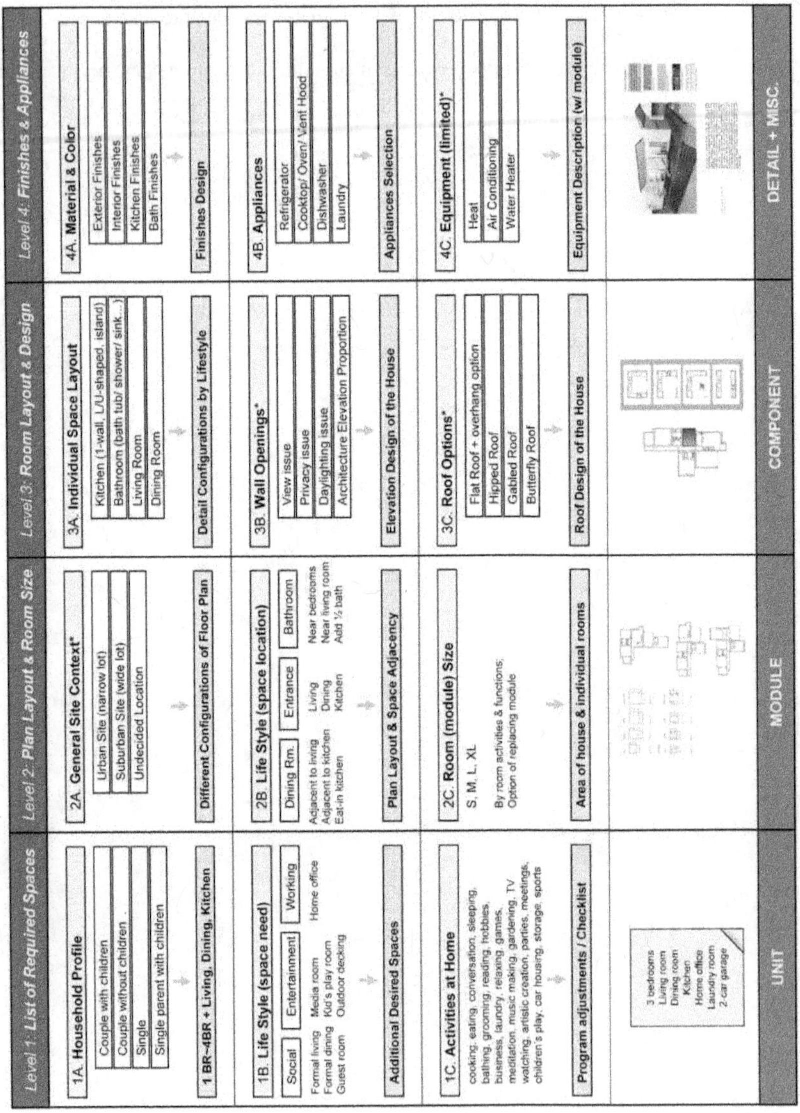

Level 1: List of Required Spaces	Level 2: Plan Layout & Room Size	Level 3: Room Layout & Design	Level 4: Finishes & Appliances
1A. Household Profile — Couple with children; Couple without children; Single; Single parent with children	**2A. General Site Context*** — Urban Site (narrow lot); Suburban Site (wide lot); Undecided Location	**3A. Individual Space Layout** — Kitchen (1-wall, L/U-shaped, island); Bathroom (bath tub/ shower/ sink...); Living Room; Dining Room	**4A. Material & Color** — Exterior Finishes; Interior Finishes; Kitchen Finishes; Bath Finishes
1 BR~4BR + Living, Dining, Kitchen	Different Configurations of Floor Plan	Detail Configurations by Lifestyle	Finishes Design
1B. Life Style (space need) — Social | Entertainment | Working; Formal living / Media room / Home office; Formal dining / Kid's play room; Guest room / Outdoor decking	**2B. Life Style (space location)** — Dining Rm. | Entrance | Bathroom; Adjacent to living / Living / Near bedrooms; Adjacent to kitchen / Dining / Near living room; Eat-in kitchen / Kitchen / Add ½ bath	**3B. Wall Openings*** — View issue; Privacy issue; Daylighting issue; Architecture Elevation Proportion	**4B. Appliances** — Refrigerator; Cooktop/ Oven/ Vent Hood; Dishwasher; Laundry
Additional/Desired Spaces	Plan Layout & Space Adjacency	Elevation Design of the House	Appliances Selection
1C. Activities at Home — cooking, eating, conversation, sleeping, bathing, grooming, reading, hobbies, business, laundry, relaxing, games, meditation, music making, gardening, TV watching, artistic creation, parties, meetings, children's play, car housing, storage, sports	**2C. Room (module) Size** — S, M, L, XL; By room activities & functions; Option of replacing module	**3C. Roof Options*** — Flat Roof + overhang option; Hipped Roof; Gabled Roof; Butterfly Roof	**4C. Equipment (limited)*** — Heat; Air Conditioning; Water Heater
Program adjustments / Checklist	Area of house & individual rooms	Roof Design of the House	Equipment Description (w/ module)
3 bedrooms; Living room; Dining room; Kitchen; Home office; Laundry room; 2-car garage			
UNIT	**MODULE**	**COMPONENT**	**DETAIL + MISC.**

Figure 4.44. Integration of Questionnaire and Modular System

4.9.1 Unit. The term "unit" here indicates housing unit which is defined as a self-contained living space. It may include a house, apartment, mobile home, group of rooms, or

113

a single room intended for occupancy as separate living quarters. Generally speaking, a housing developer creates different layouts of floor plans in multi-family housing as different "unit types". The systematic approach of unit research in the Figure 4.44 first column demonstrates the potential of the digital questionnaire to generate a list of required spaces. The content of this level produces the phase of architecture programming, and does not show any design geometry to avoid a misleading first impression. The goal of these "unit-related" questions is trying to determine the client's needs rather than having the client select images from the database of a design gallery.

4.9.2 Module. The module can be a spatial module or a physical module. A spatial module may or may not as an independent module for shipping to the construction site. A spatial module can be a graphical representation of a "room" in the consumer-driven web-based interface for the client to easily add, remove, or replace the different spaces in the plan layout within the database options. There are three sub-categories under the Level 2 Questionnaire (Figure 4.44 second column): A. General Site Context, B. Life Style, and C. Room Size. Collecting data of site context is important because it can avoid the problem that a suggested design option does not fit the site boundary (i.e. the house is too big for the urban narrow lot). However, the issue of site can become a very complex category if it involves grading, drainage, landscaping, etc. The questionnaire of this system simplifies this issue by using three options: urban site (narrow lot), suburban site (wide lot), and undecided location (show both conditions).

The life style of the Level 2 Questionnaire aims to define the space location rather than the space need in the Level 1 Questionnaire. For example, by defining the space adjacent preference of dining area, entrance, and bathroom location, the database can generate possible plan layouts from the recommended space program list.

Besides the plan layout, the size of room reflects the different activities and functionalities of each space and will probably not be the same size for all. Except for a linear organization of plan layout, changing any room size may affect other adjacent room's dimension or the circulation of the house. Figure 4.45 is an example of different room dimensions of the same square proportion for (a) master bedroom, (b) living room, (c) dining room, and (d) home office.

Figure 4.45. Different Room Sizes of the Same Proportion

4.9.3 Component.   The component category of this research means the individual room layout plus the interchangeable "infill" components like floor, wall and roof. Results from the Level 2 Questionnaire already define the configuration of the floor plan and the size of individual rooms. The task of the Level 3 Questionnaire is to provide the alternative content

115

of individual spaces. For example, the cooking style and the frequency of kitchen usage may determine the layout of the kitchen. A range of activities will have contributed to the reconfiguration of the kitchen design because kitchen is the most functional and possibly the most active space of a house.

Over the years, kitchen cabinetry has evolved from custom-made to modular production. A finished kitchen shell includes electricity, plumbing, and flooring. The cabinets can be ordered from a web-based catalog and shipped as preassembled modules to the site, where they are mounted on the walls, leaving space for appliances to be delivered and plugged in. This mode of construction simplifies the process from the conventional custom-made products by the carpenter or cabinet maker (Friedman, 2002). Figure 4.46 shows an example of the variety of kitchen layouts that can be designed to reflect the different life styles and space limitations of a house. All layouts can be considered as different kitchen modules which leave exchangeable space for kitchen appliances to plug in and plug out.

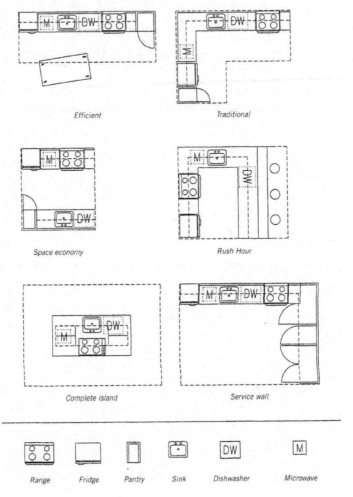

Efficient

Traditional

Space economy

Rush Hour

Complete island

Service wall

Range   Fridge   Pantry   Sink   Dishwasher   Microwave

Figure 4.46. A Variety of Kitchen Layouts

Besides the interior room layout, the component also includes wall openings and a roof option. Wall openings can be represented within a design palette combining different door and window configurations based on the dimensions of a modular system. Figure 4.47 illustrates façade options that utilize the palette of door and window options with a 7' wide and 8' high infill module (Delary, 2004).

Figure 4.47. Palette of Door and Window Options

Generally speaking, there are four different types of roof: flat roof, gabled (pitched) roof, hipped roof, and butterfly roof in residential architecture. Each of the terms can also be subdivided to many different styles by the historical period of development. In modern prefab houses[18], there are fewer decorative elements and more of an expression of volumetric simplicity. For instance, the roof option provides by the weeHouse vender has only one type – flat roof. There are two choices under the flat roof type: no overhang or 3' overhang (Figure 4.48)

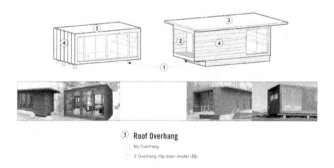

③ Roof Overhang
　No Overhang
　3' Overhang (flip down onsite) ($$)

Figure 4.48. Example of Roof Options

4.9.4　Detail.　The term "detail" here is used to cover miscellaneous items that were not included in pervious categories. However, the major items will be: finishes and appliances. First, the finishes indicate any surface material or color from exterior and interior building

---

[18] Modernist Prefab Dwellings, http://www.fabprefab.com/

components. It can also be sub-divided into any individual space like kitchen finishes or bath finishes for further customization. Figure 4.49 is an example of choosing different exterior finishes of a prefabricated modular house provided by the m-house vendor from London. Toyota also provides a customizable façade option from their Toyota Home's website (Figure 4.50).

Figure 4.49. Example of Customizing Exterior Finishes from m-house

Figure 4.50. Example of Customizing Exterior Finishes from Toyota Home

Appliances may include the kitchen appliances, laundry appliances, and plumbing fixtures. Unlike furniture, appliances usually need to hook up with pre-determined electrical and plumbing outlets. As mentioned in Chapter 4.9.3, a kitchen module can include a lifestyle fitted kitchen layout with a combination of cabinet, cooktop, oven, dishwasher, refrigerator, etc. Figure 4.51 is an example of selecting different kitchen appliances from the Marmol Radziner Prefab modular house vendor's website.

Figure 4.51. Example of Selecting Kitchen Appliances from Marmol Radziner Prefab

4.10   Interface Design and Web Organization

The proposed i_Prefab Home design interface for customizing prefabricated housing shows in Figure 4.52. When a user puts the correct webpage address in the internet browser, a static HTML page will open with a welcome message and simple instructions for this intuitive system. The web page has been divided into six frames; besides the information and design process frames on the top plus view navigation and output services frames on the bottom of the interface for different purposes, the system actually only contains two main components: client's input on the right window and design visualization output on the left window.

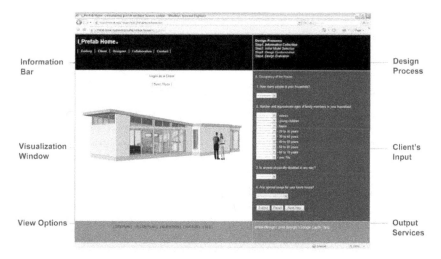

Figure 4.52. i_Prefab Home Design Interface

Below is the outline of the six elements in the proposed design interface:

(1)   Information Bar/ Navigation Window: includes login for client and professional, collaboration method, contact information, and design gallery.

(2)   Design Process Window: indicates the sequence and status of current step.

(3)   Visualization Window: shows 2D/ 3D images of design suggestion and customizable items in each important step.

(4)   Client's Input Window: questionnaire by text description and graphics selection.

(5)   View Options Window: provides different building or component views to understand and visualize the design.

(6)   Output Service Window: shows the options to deliver the end result.

Besides the web-based design interface, the system is proposed to be used by:

(1)   The client alone with instant assist from online help center.

(2)   An architect with the client.

(3)   An architectural-trained salesperson with the client.

Figure 4.53 shows the general web structure and major consulting procedures. The navigation window on the top-left has four main categories: Gallery, Client (basic mode), Designer (professional mode), and Collaboration. The consulting procedures represent the different phases of architectural programming and highlight the current status in the Design Process Window. In this prototype simulation, only the consumer-driven basic mode will be implemented along with the related information contents. Future research will be described in the last chapter that the supply chain collaboration by online digital BIM models & available technologies will be integrated into a complete and comprehensive system.

The website organization diagram introduces six important steps of the design consulting process, and five of them can be achieved in the basic mode with fully participated by the client. At the end of each step, the system can provide a preview of the current result as a temporary reference, and at anytime during the process, the user can go back to the previous step for changing the selected options. The optimum goal of the prototype system is to record the client input data and output design results as the future reference for anyone who matched the similar input data may get the design suggestion form the database. Currently, the Amazon.com website already implemented this matching favorite concept. With the matured development of artificial intelligent (AI) technology in the near future, machine-learning capability is not just as a dream.

In the next chapter, detailed procedures of the web-based, consumer-driven advisory system for modular house design will be discussed with a hypothetical client, as well as the methods of professional help and evaluation.

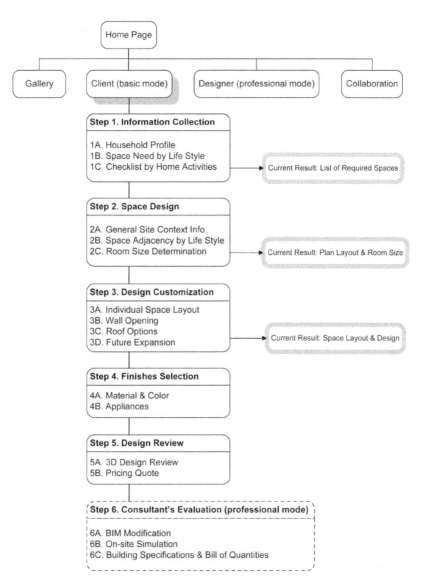

Figure 4.53. Website Organization of the Prototype Simulation

CHAPTER 5

CASE STUDIES AND PROTOTYPE SIMULATIONS

5.1   Case Study Overview

In order to show how this proposed model works, a prototype simulation is developed

to provide the modular housing design consulting and customization for the consumer as a

simplified web-based application in the early phase of the housing delivery process. A series

of new concepts and methods like decision tree visualization, digital questionnaire as the data

collection, and different levels of design suggestion have been integrated as a prototype to

apply in real cases. At each step of this consulting process, there is a screenshot showing the

input/ output simulation results. The text information on the right window of the interface

includes questionnaire, explanation and multiple choices for client to input (Figure 5.1). The

graphic information on the left window is the visualization output from the advisory system.

The information window on the upper-right corner indicates the status of design process.

Overall, the intent of this interface design is to try to keep simple and intuitive.

In order to demonstrate the prototype of the i_Prefab advisory system, two

hypothetical client conditions have been specified for two-bedroom and four-bedroom case

studies to represent household profile, life style, and preference selection.   Strategic

information about the client preferences is obtained through a series of digital questionnaire,

which are then interpreted into visualizations as customizable design options. Clients are

encouraged to experiment with changing preferences in order to better visualize trade-off and

compromises.

The purpose of these two case studies intends to be extremely different conditions –

one is an economical two-bedroom house in an urban narrow lot, the other one is a luxury

four-bedroom house in a suburban site type. The results of case studies can be applied for the

future research since most living cases are somewhere in between.

125

5.2   Case Study One – Two-bedroom House

The client is a retired couple that decided to move from suburban to an urban residential area in order to take advantage of city life.   Unlike suburban, urban life provides more social activities like museum exhibitions, music concerts, and street window shopping, etc. The client is planning to use the i_Prefab design consulting tool to see the possibility of tailoring their home to their needs at an affordable price without the benefit of a full service architect.   The client has already purchased land, a narrow lot (25'x125') with front yard and back yard.

5.2.1   Level 1: List of Required Spaces.   The first page is an introduction to provide an overview of the i_Prefab advisory system and basic instructions for creating a personal account.   After the client creates an account, the first step of the design interface is the gathering the client's information to generate a list of required spaces. There are two tasks in the information gathering step: A. Household Profile and B. Life Style.

The first question tries to define the household profile in the new home (Figure 5.1). The household profile here indicates the major occupants for the new house and does not relate to family profile. Therefore, the client selects "couple without children" to represent their current household profile.

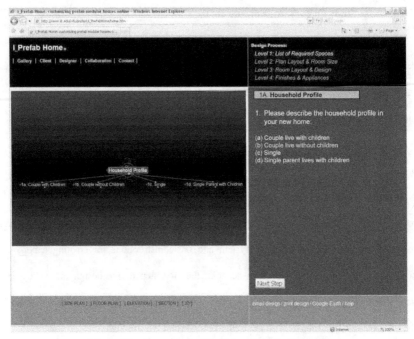

Figure 5.1. Screenshot of Defining Household Profile

The content of the question will be based upon the answer of the preceding question. The client selects "older couple live as empty nesters" to describe their case (Figure 5.2). In some cases, this question will branch to queries regarding future room requirement needs. For instance, a young couple planning to have children will affect the number of bedrooms in the list of required spaces.

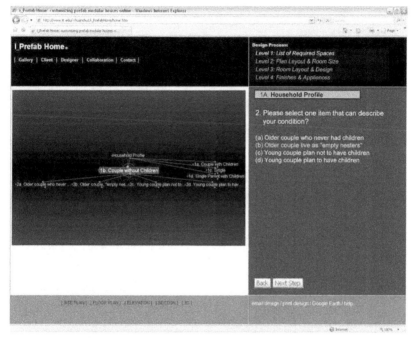

Figure 5.2. Screenshot of Defining Household Profile

The third question is based on the selection of "older couple live as empty nesters" to determine if there is a requirement of having a guest room. The retired couple (client) chose the option of planning to have a guest room, because they are willing to accommodate overnight guests, especially when any of their children visits. Question No. 4 considers the requirements of ADA bathroom and ramps if anyone is physically handicapped. Figure 5.3 shows that the decision-tree navigation diagram in the visualization window highlights the current status of the information input process as well as representing the relationship with the other options. After answering four questions above, the i_Prefab Home system analyzed that this household profile requires minimum of two bedrooms. The Tree Diagram in Figure 5.4 illustrates possible living unit suggestions based on the different household profiles in the design process step 1A.

128

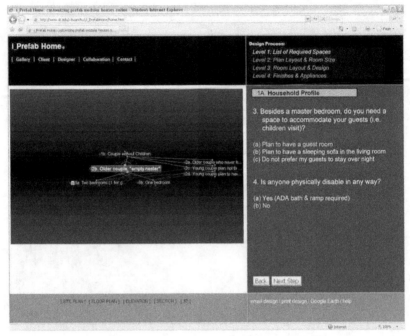

Figure 5.3. Screenshot of Defining Household Profile

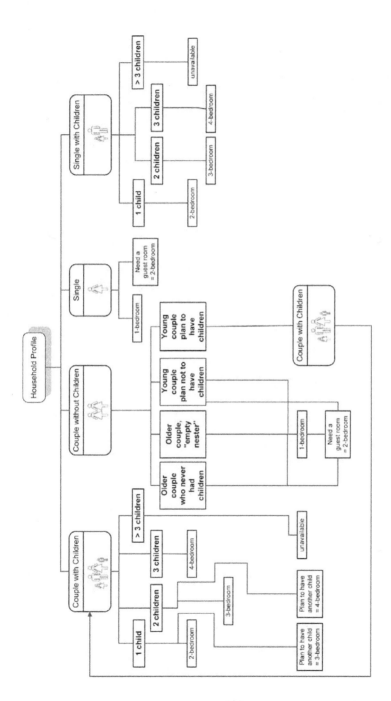

Figure 5.4. Tree Diagram of Selecting Living Units

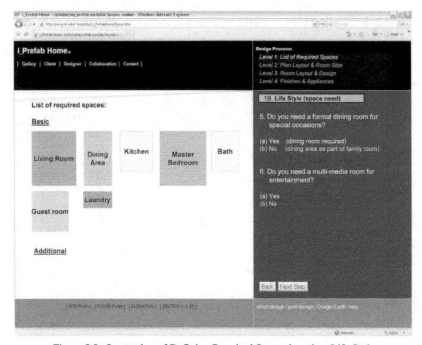

Figure 5.5. Screenshot of Defining Required Spaces based on Life Style

Questions No. 5 to No. 7 define the specific space requirement based on the client's lifestyle. Question No. 5 asks the requirement of a formal dining room for special occasions. If the client answered no, then a dining area will be provided as a part of the family room. In the meantime, the visualization window will provide diagrammatic blocks for all required spaces (Figure 5.5). The square footage of the space block indicates the available size of the prefabricated components that have been analyzed in Figure 4.19 of Chapter 4.6. Economically, if the client can be satisfied with the pre-made modules without any major customization, they will save a lot of time and money.

Question No. 6 addresses the requirement for a multi-media room for entertainment, and question No. 7 defines the need for a home office. In most cases, the i_Prefab system provides three different options to match the spatial representation. We can use IF/THEN logic to explain Figure 5.6.

IF variable is *set* (a) need a formal home office, THEN *action* home office option.

IF variable is *set* (b) need an informal working area to work sometimes, THEN *action* den option.

IF variable is *set* (c) never work at home, THEN *action* no home office need.

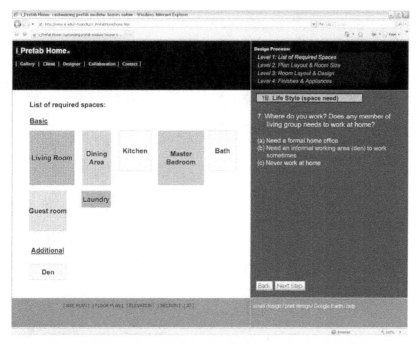

Figure 5.6. Screenshot of Defining Required Spaces based on Life Style

5.2.2   Level 2: Plan Layout and Room Size.   Compared to Level 1 design process, Level 2 Questionnaire is more focused on the building geometry and space adjacency. There are two sub-categories under Level 2 design process: General Site Context and Life Style. As mentioned previously, collecting data of site context is important because it can avoid the problem that a suggested design option does not fit the site boundary. The first question in

132

this phase is to gather the general site context information by three simplified options: urban residential lot, suburban residential lot, and rural site. The visualization window uses aerial photos to provide a clear representation of different site contexts. Figure 5.7 shows the client selected the urban residential lot as the future house location. Once the client selected the urban residential lot, there is another follow-up question to define the relationship between the lot geometry and adjacent street. Question No. 9 addresses four different conditions of urban residential lot: (a) single and inside lot, (b) single and corner lot, (c) double and inside lot, and (d) double and corner lot (Figure 5.8).

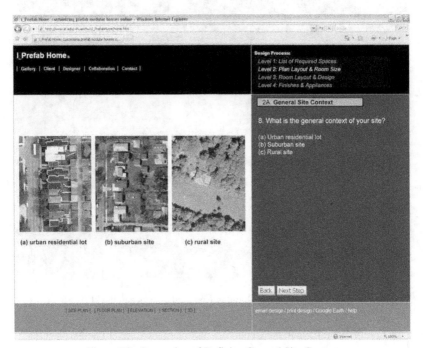

Figure 5.7. Screenshot of Defining General Site Context

Figure 5.8. Screenshot of Defining General Site Context

The questions followed by the general site context are related to the issues of view, privacy and space adjacency. Questions No. 10 and No. 11 ask the importance of the surrounding views and the best view's direction (Figure 5.9). In order to determine the first room module's location, question No. 12 defines the first space fronting on the street regarding the privacy and lifestyle issues (Figure 5.10). Once the system has the first module located as a reference point, the rest of the modules can be easier to locate based on the room adjacency preference. The visualization window in Figure 5.11 shows four design suggestions which are suitable for the standard urban lot condition from the database. Figure 5.12 and Figure 5.13 explain the logic of floor plan selection based on the site constraint.

134

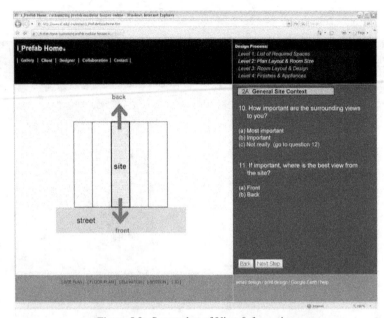

Figure 5.9. Screenshot of View Information

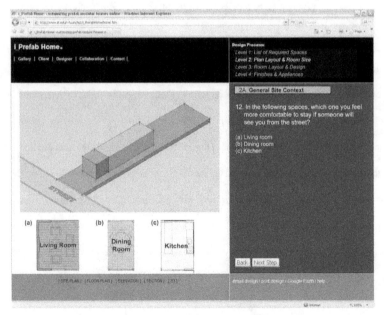

Figure 5.10. Screenshot of Defining the First Space

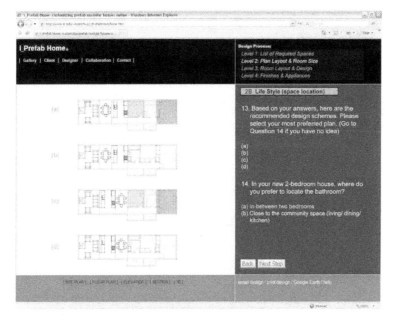

Figure 5.11. Screenshot of Defining Space Adjacency

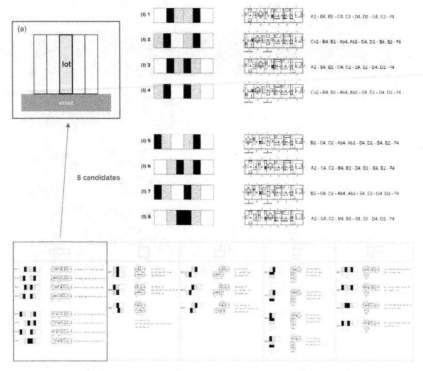

Figure 5.12. Logic of Candidate Design Selection

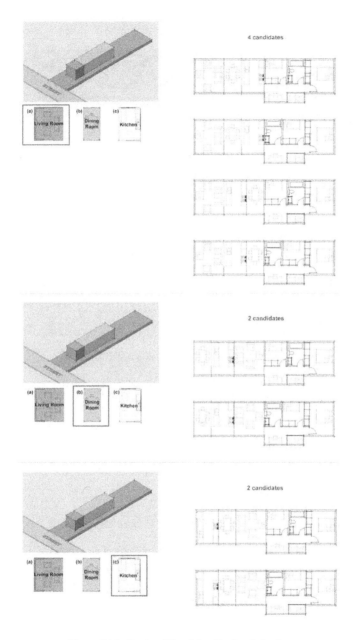

Figure 5.13. Logic of Candidate Design Selection

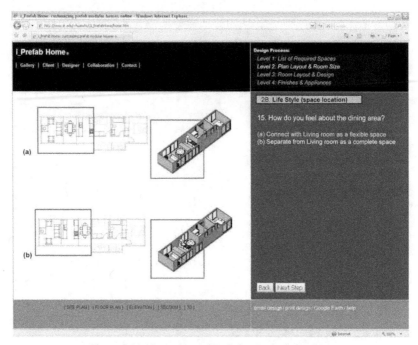

Figure 5.14. Screenshot of Defining Space Adjacency

If the client is unsure of the most preferred plan of these four design suggestions, he or she can skip question No. 13. Answering the next question, No. 14 related to bathroom's location narrows down the design choices to two options. Figure 5.14 shows the 3D views of design options for dining and kitchen space arrangements provides a clear spatial feeling for the user's to make a judgment. Question No. 16 asks the first interior image impression for defining the location of the entrance (Figure 5.15). After answering all "Level 2 – Plan Layout and Room Size" questions, Figure 5.16 is the confirmation of current design phase.

Figure 5.15. Screenshot of Defining the Entrance

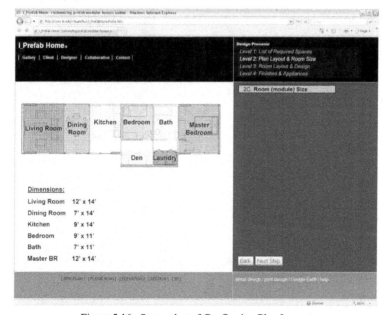

Figure 5.16. Screenshot of Confirming Plan Layout

140

5.2.3 Level 3: Room Layout and Design. The building geometry and floor plan boundary have been decided before moving to this phase of design process. The task of this phase is to provide the alternative content of individual spaces. The options of different room layout have been treated as replaceable and compatible "space tiles". The four different kitchen layouts in Figure 5.17 are examples for showing the graphic layout in the visualization window and feature explanations in the client's input window. The different kitchen layouts may reflect the different cooking and living style of the clients. A cost indicator under the optional layouts provides the idea of price and can output a constant update of the cumulative financial ramifications for making decisions.

Figure 5.17. Screenshot of Selecting Individual Space Layout

The relationship between the questionnaire, answer, and spatial conversion is illustrated in Appendix A. The explanation here is simplified by IF/THEN rules:

141

For Question No. 17:

IF variable is *set* (a) Most of time, THEN *action* - More countertop space.

IF variable is *set* (b) Sometimes, THEN *action* - Standard (Default setting).

IF variable is *set* (c) Rarely, THEN *action* - Less countertop space.

For Question No. 18:

IF variable is *set* (a) Family meals, THEN *action* - Large size kitchen.

IF variable is *set* (b) Quick & simple meals, THEN *action* - Medium size kitchen.

IF variable is *set* (c) Microwave food, THEN *action* - Small size kitchen.

For Question No. 19:

IF variable is *set* (a) No one else in the kitchen while preparing meals, THEN *action* - Gallery or single bar kitchen layout.

IF variable is *set* (b) A helper in the kitchen when preparing meals, THEN *action* - L-shape kitchen layout.

IF variable is *set* (c) Family or friends visiting during meal preparation, THEN *action* - L-shape w/ island kitchen layout.

Similar to the kitchen layout, the dining space has two variables of dining table size and the spatial relationship with the adjacent spaces. The size of the dining area and the dining table determine how many people can have a dinner together in the same space. Since the dining table is not fixed furniture, the client can always change it later. The question in Figure 5.18 addresses the spatial relationship of the dining area and its adjacent spaces. The graphical diagram shows there are three different options as (a) open to the adjacent spaces, (b) enclosed dining room, and (c) open to the living room and has a small opening on the adjacent wall for the food delivery. This example demonstrates although the dining space is

142

an independent "fully modular[19]" part, it still can be sub-divided as a component-based

module for further customization of its side walls.

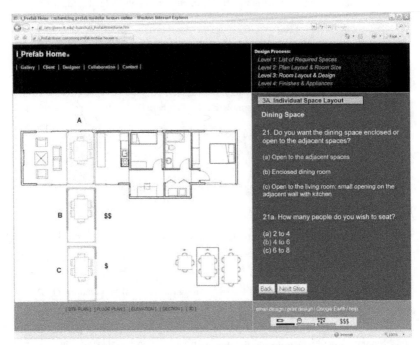

Figure 5.18. Screenshot of Selecting Individual Space Layout

After the individual space layout, the visualization window of the user interface

moves from the two-dimensional floor plan view to the three-dimensional isometric building

view in order to show the customizable options of wall openings and roof types. Based on

the demonstration of a 2-bedroom modular house example in Figure 5.19, there are two

choices of the wall openings: (a) "open" elevation with blinds and (b) "closed" elevation with

---

[19] See the definition of System Types by Basic Modular Element in Figure 2.12, Chapter 2.3.

clerestory. Further explanation of both pro and con are provided in each option for the client to make a decision.

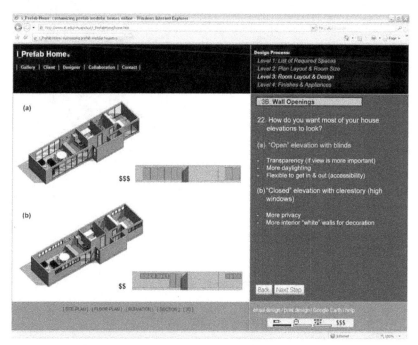

Figure 5.19. Screenshot of Selecting Wall Opening Options

In addition to wall openings, the options of roof allow the client to customize the roof type beyond the "cookie-cutter" uniformity of prefabricated housing in the past. The advantages have been described in each option, and the simplified options here can be expanded based on the local climate condition. Figure 5.20 shows the hypothetical client selected (b) single-slope roof as the preference.

144

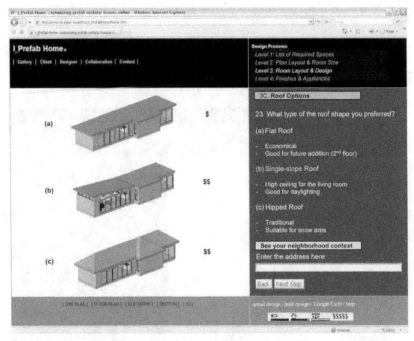

Figure 5.20. Screenshot of Selecting Roof Options

The next step after roof options is called future expansion. It is important to be concerned about the future expansion because the current occupants of the proposed house are only two people. The configuration of household profile may be changed after a couple years (i.e. divorced daughter moved back to live with parents). A good strategy of the adaptability of homes is to integrate future expansion option into the design before construction begins (Friedman, 2002). Figure 5.21 demonstrates two expansion options: vertical and horizontal with the precise modular components and price for client as a reference.

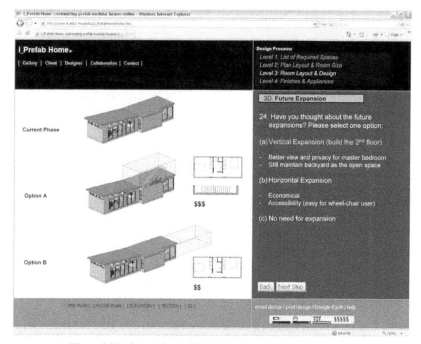

Figure 5.21. Screenshot of Selecting Future Expansion Options

146

5.2.4 Level 4: Finishes and Appliances.   Once the building geometry, wall openings, and roof forms have been decided, the rest of the customization options are about the finishes and appliances. First of all, the finishes indicate any surface material or color for exterior and interior building components. The concept of customizing building elevation materials or interior finishes have been proposed and even implemented by existing modular housing vendors. The reason to include this method is to prove that the i_Prefab advisory system can achieve this level and even better. Figure 5.22 shows the proposed modular house design model has been adjusted to human scale perspective as a normal view. Exterior siding materials and colors can be customized by a simple mouse click. Scaled people are provided as reference.

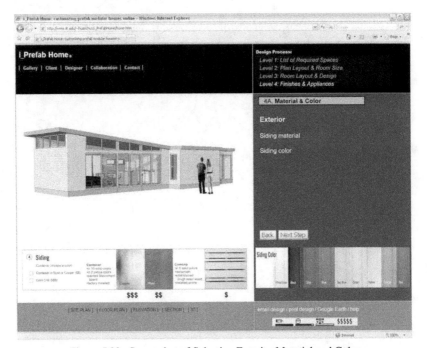

Figure 5.22. Screenshot of Selecting Exterior Material and Color

Moving from the exterior to the interior, each room has a simple perspective with ID numbers to identify its space element, like floors, walls, ceiling, doors, trim, curtain track, etc (Figure 5.23). After the client selects the desired finishes, the interior perspective will show as a real-time rendering.

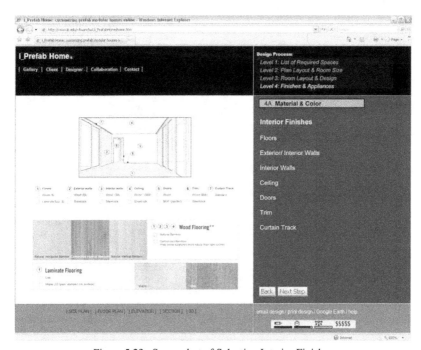

Figure 5.23. Screenshot of Selecting Interior Finishes

Kitchen and bathroom have the most built-in furniture. The customization process of these rooms requires a more detailed level. For example, kitchen finishes include many components: doors, countertops, cabinets, and kitchen accessories (Figure 5.24). Bathroom finishes may include floors (wood, laminate, tile), walls (tile, partial tile, wood, sheetrock), and ceiling (wood, sheetrock).

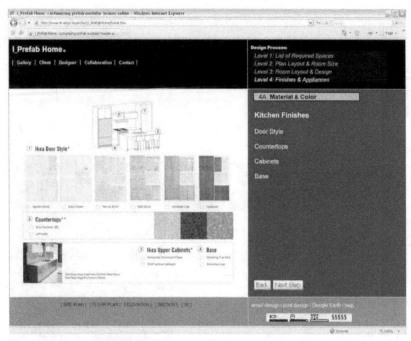

Figure 5.24. Screenshot of Selecting Kitchen Finishes

The last but not the least is the appliances selection process. Appliances for a residential project may include kitchen appliances and laundry appliances. There are many different options of these appliances in the market. However, the sizes are varied and even the same functional product has many different sizes. In order to coordinate with the other interior components, i.e. kitchen countertop, gathering a precise dimension is necessary before the construction process. Figure 5.25 displays the available kitchen appliances that fit into this modular system with a green highlight indicated in the floor plan.

149

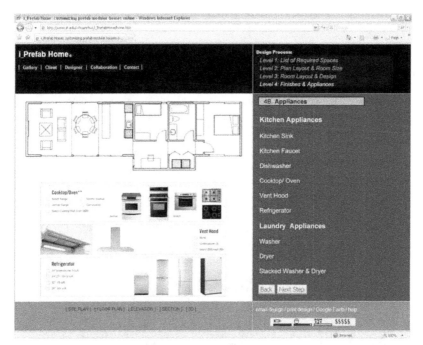

Figure 5.25. Screenshot of Selecting Kitchen Appliances

5.3   Case Study Two – Four-bedroom House

The second case study is a four-bedroom house unit located in a suburban site condition. This case represents most middle-class families who prefer to live in a suburban setting to provide a better environment for children to play and study, especially true in North America. A couple with two children are planning to settle in a suburban area. The client uses the i_Prefab design consulting tool to see the possibility of tailoring their single family detached house to their needs at an affordable price for the schematic design phase. The client has already purchased land as a suburban residential lot with the dimension of 100'x125' (~0.28 acre).

5.3.1   Level 1: List of Required Spaces.   As Chapter 5.2.1 mentioned, the first page is an introduction to provide a big picture of the i_Prefab advisory system and basic instruction for creating a personal account. After the client creates an account, the first step of the design

interface is the gathering the client's information to generate a list of required spaces. There

are two tasks in the information gathering step: A. Household Profile and B. Life Style.

The first question tries to define the household profile in the new home (Figure 5.26).

The household profile here indicates the major occupants for the new house and does not

relate to family profile. Therefore, the client selects "couple live with children" to represent

their current household profile.

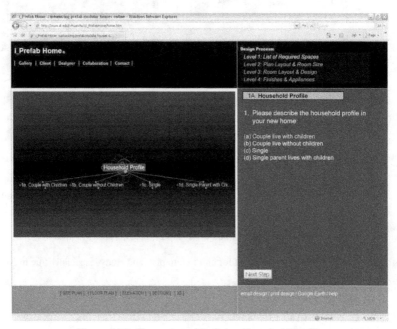

Figure 5.26. Screenshot of Defining Household Profile

The content of the question will be based upon the answer of the preceding question.

The second "follow-up" question is "How many children do you have?" The client selects

"Two" to describe their case (Figure 5.27). In some cases, this question tries to determine the

future needs of the room requirements. For instance, the question No. 3 asks "Do you anticipate having another child within five years?" (Figure 5.28). If the couple is planning to have a third child, the system will change the number of bedrooms in the suggested list of required spaces.

Figure 5.27. Screenshot of Defining Household Profile

Question No. 4 collects the information of children's age and gender in order to determine the room arrangement. Figure 5.28 shows that the decision-tree navigation diagram in the visualization window highlights the current status of information input process as well as representing the relationship with the other options. After answering four questions above, the i_Prefab Home system analyzed this household profile and found it is suitable to have a minimum of three bedrooms (Refer to Figure 5.4).

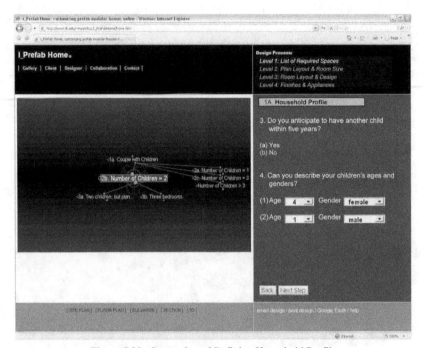

Figure 5.28. Screenshot of Defining Household Profile

The questions from No. 5 to No. 10 define the specific space requirement based on the client's lifestyle. Question No. 5 defines the need of a home office. Using indirect questions of "Where do you work?" and "Does any member of living group needs to work at home?" is a strategy of reflecting the client's lifestyle. Question No. 6 asks the requirement of a formal dining room for special occasions. If the client answered no, then a dining area will be provided as a part of the family room. In the meantime, the visualization window will provide diagrammatic blocks with approximate square footage for all required spaces (Figure 5.29).

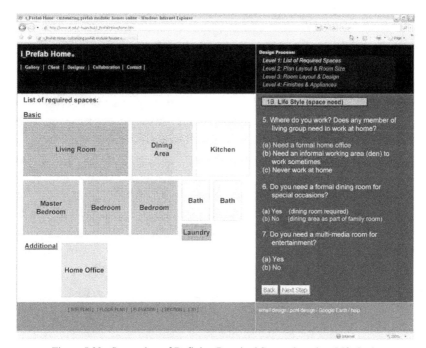

Figure 5.29. Screenshot of Defining Required Spaces based on Life Style

Questions No. 7 addresses the requirement of a multi-media room for entertainment, and question No. 8 defines the need of a guest room or just converts the living room with a sleeping couch as a temporary guest room. Based on the answer of question No. 8, the client selected (a) plan to have a guest room. In the meantime, the question No. 9 is a follow-up question – "Based on the previous question, can home office double as a guest room?" If the client answers yes, the space for home office can also be used as guest room once in a while. The "space block" in the visualization window will become a dashed boundary to represent that one space occupies two different functions (Figure 5.30). Question No. 10 is the last question in the Level 1 Questionnaire regarding the need of basement. Basement can be used for laundry room, storage, and children's playroom. In northern climates, houses usually are built with crawls spaces. To create a basement can be accomplished at relatively low cost.

154

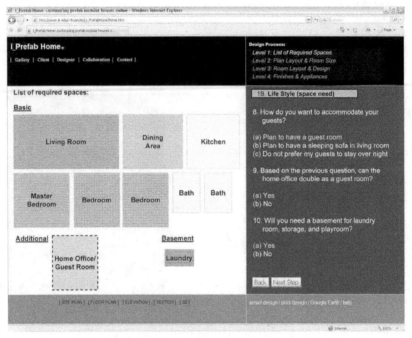

Figure 5.30. Screenshot of Defining Required Spaces based on Life Style

5.3.2 Level 2: Plan Layout with Site Context.   Compared to Level 1 design process, Level 2 Questionnaire is more focused on the building geometry and space adjacency. There are three sub-categories under Level 2 design process: General Site Context, Life Style, and Adaptability. As mentioned previously, collecting data of site context is important because it can avoid the problem that a suggested design option does not fit the site boundary. The first question in this phase is to gather the general site context information by three simplified options: urban residential lot, suburban residential lot, and rural site. The visualization window uses aerial photos to provide a clear representation of different site types. Figure 5.31 shows the client selected the suburban site as the future house location. Once the client selected the suburban residential lot, there is another follow-up question to define the relationship between the lot and adjacent street. Question No. 12 addresses four different

155

conditions of urban residential lot: (a) narrow side facing street, (b) corner lot, (c) wide side

facing street, and (d) Cul-de-sac (Figure 5.32).

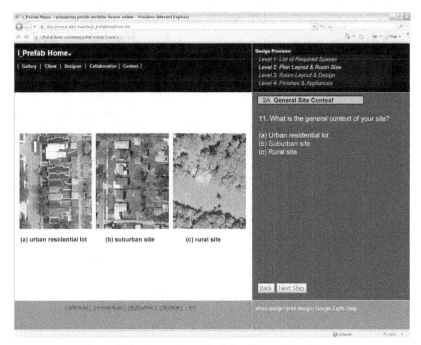

Figure 5.31. Screenshot of Defining General Site Context

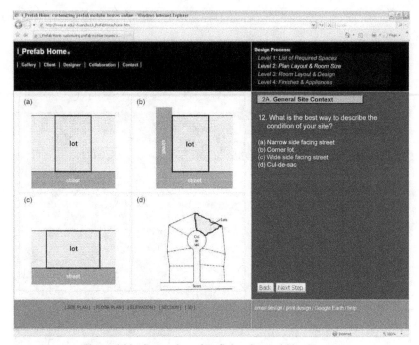

Figure 5.32. Screenshot of Defining General Site Context

The questions followed by the general site type are related with the issues of view, privacy and space adjacency. Question No. 13 and Question No. 14 ask the importance of the surrounding views and the best view's direction. In order to determine the vertical relationship of space adjacency and receive precise suggestions from the database, question No. 15 asks the preferred number of stories in the new house (Figure 5.33).

157

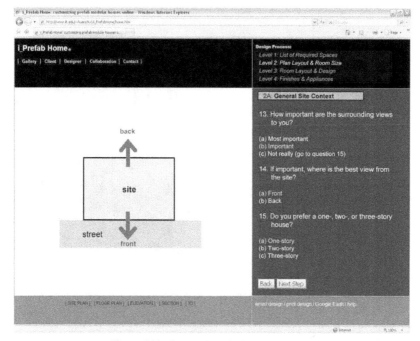

Figure 5.33. Screenshot of View Information

Different than the two-bedroom approach, the complexity level is much higher in the four-bedroom space planning due to the increased alternatives of the program and greater flexibility of residential lot. Question No. 16 intends to organize the "public space" modules by asking the preference of open floor plan or a traditional divided floor plan for the living room, dining room, and kitchen (Figure 5.34). Figure 5.35 shows two different proportions of the living room/dining/ kitchen. One provides a depth of interior space with a long wall, and the other one has a square plan for flexible usage.

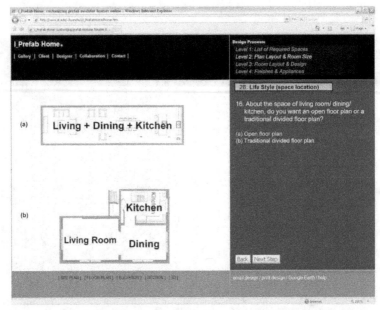

Figure 5.34. Screenshot of Defining the Space Modular Group

Figure 5.35. Screenshot of Defining the Space Proportion

Questions No. 18 to No.20 are related to the space adjacency issue. Similar to the space adjacency matrix in the architectural programming process, these three questions intends to define the three relationships: (a) adjacency immediate, (b) no comment, and (c) avoid adjacency between the selected two spaces. For the convenient purpose, the client selected home office/guest room to connect immediately with the living/dining/kitchen space module. For the privacy issue, the client prefers to locate their master bedroom away from the living/dining/kitchen space module and home office/guest room (Figure 5.36).

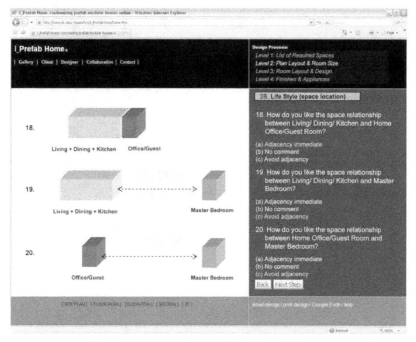

Figure 5.36. Screenshot of Defining Space Adjacency

Once the client answers the space adjacency questions, the i_Prefab system can generate the best fit of four-bedroom layouts based on the customer's inputs. The visualization window in Figure 5.37 shows two design suggestions – one is like an asymmetric U-shape with a courtyard, and the other one is a T-shaped plan. Until this point,

160

the system has narrowed down from many choices to two options for the client to work on further details. Figure 5.38 and Figure 5.39 explain the logic of candidate design selection based on the required program, number of floor, space adjacency, and proportion issues.

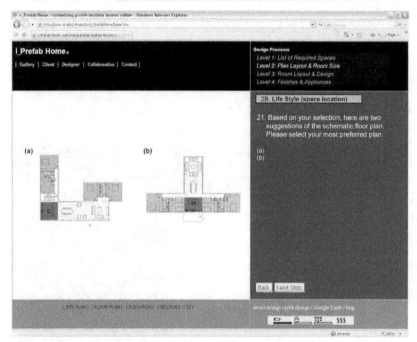

Figure 5.37. Screenshot of Floor Plan Suggestions

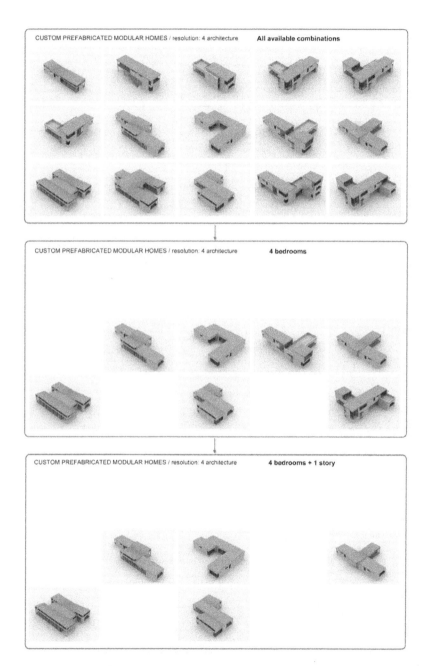

Figure 5.38. Logic of Candidate Design Selection

162

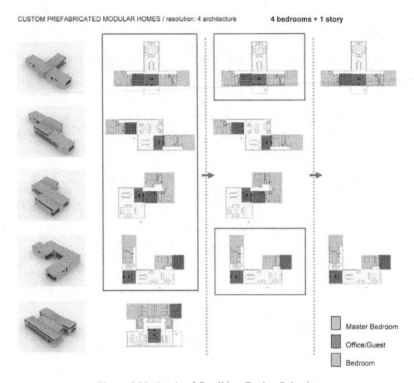

Figure 5.39. Logic of Candidate Design Selection

After the floor plan layout, questions No. 22 and No. 23 are intended to determine the location and size of the garage (Figure 5.40). Question No. 22 asks the preference of attached garage or detached garage. Question No. 23 defines the number of cars that the client's family members drive regularly. Considering the market resale value, the system provides at least two-car garage even if the client has only one car. There is a complete sequence of the prototype questionnaire in Appendix B for reference.

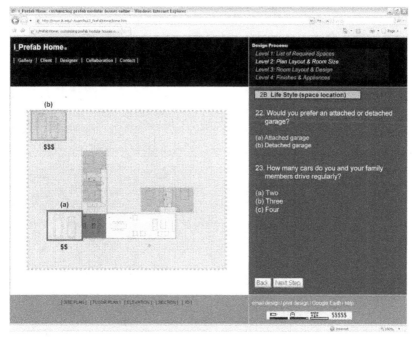

Figure 5.40. Screenshot of Defining the Garage

The category 2C is adaptability suggestions. In Figure 5.41, the floor plans in the visualization window present the current room arrangement (above) and future room arrangement (below). Since the client's two children are only as 4-year-old girl and 1-year-old boy, the children's bedroom will be convenient if located close to the master bedroom, and leave the guest room and home office independently on the other side of the house. Once the children grow up, these rooms can switch back to provide two independent rooms for girl and boy, and move the home office/guest room close to the master bedroom.

164

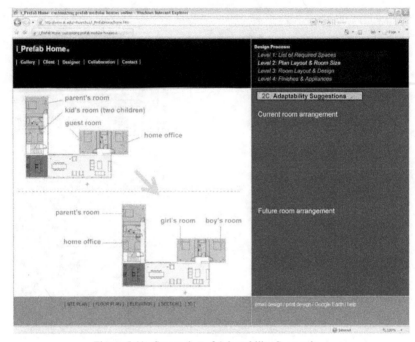

Figure 5.41. Screenshot of Adaptability Suggestions

5.4   Professional Remote Help and Design Center

Based on the previous discussion of Dr. Noguchi's Mass Custom Home Model (Noguchi, 2003) in the literature review, a good model of integrating mass customization concept is not only dealing with the product, but also caring about the service. He illustrates that the fundamental service factors of mass customized modular housing that may be involved with location, personnel, and the customizing tools.   In the proposed i_Prefab system, the "tool" means the web-based advisory tool to convert the client's needs to the spatial representation and design suggestions. The "location" of providing service may be delivered from the client-driven website anywhere with internet connections or there could be the physical location of a design center with a professional representative's help.   The

165

"personnel" means professional's help for assisting the clients. This section will focus on the degree and method of the potential involvement from a professional.

Figure 5.42 shows the different involving degrees of professional help. The X-axis indicates the communication behavior from a passive suggestion to a positive mentoring. The Y-axis points out the characteristics of human interaction from a virtual meeting (remote) to a physical (face-to-face) meeting. Besides the non-interactive webpage of Frequently Asked Questions (FAQ), there are three different levels of virtual remote help: instant messaging, virtual meeting, and computer screen remote controlling. For optimum communication with real people, face-to-face consulting exists in a Housing Design Center with a professional salesperson's help, though there may be some limitations of time and location. Furthermore, the technology of Virtual Reality enables a user to perform operations on the simulated system and shows the effects in real time. It has a lot of potential to be integrated as part of the design center facility in the future. There are more details of these professional help in the following sections.

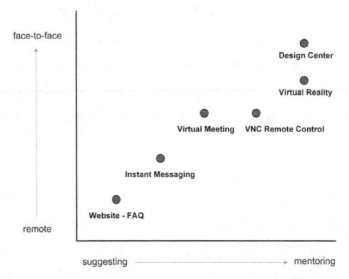

Figure 5.42. Different Involving Degrees of Professional Help

5.4.1 Frequently Asked Questions (FAQ).   FAQ is an abbreviation for "Frequently Asked Questions". The term refers to listed questions and answers, all supposed to be frequently asked in some context and pertaining to a particular topic. This is the simplest way to answer questions about problems and generate some ideas by the users themselves. The website administrator may frequently update the content of FAQ to keep it current and intelligent.

If clients try to solve the technical problems experienced from the website navigation, FAQ is not the best solution. In most cases, the FAQ webpage of a customizable product's website shows many issues regarding the product's specialty, availability, and pricing issues rather than the trouble-shooting of technical problems from the interface itself (Figure 5.43). In other words, the "trouble-shooting" FAQ is more like the "Help" in any professional software applications, with an index of issues and a keyword searching field.

Figure 5.43. Example of FAQ Webpage from weeHouse Vendor

168

5.4.2 Instant Messaging. Instant Messaging (IM) is a form of real-time communication between two or more people based on typed text[20]. The text is conveyed via computers connected over a network such as the Internet. Instant messaging requires the use of a client program that hooks up to an instant messaging service and differs from e-mail in that conversations are then possible to happen in real time. Most services offer a presence information feature, indicating whether people on one's list of contacts are currently online and available to chat. This may be called a contact list. Popular instant messaging services on the public Internet include Windows Live Messenger, AOL Instant Messenger, Excite/Pal, Gadu-Gadu, Google Talk, iChat, ICQ, Jabber, Qnext, QQ, Meetro, Skype, Trillian, Yahoo! Messenger and Rediff Bol Instant Messenger. In order to chat with the other person, both computers need to have the same Instant Messaging software installed, and the parties involved with the communication must be currently online.

Figure 5.44 is an example of utilizing Windows Live Messenger to help the client when they have hesitated in answering a question in the i_Prefab advisory tool. The advantage of Instant Messaging is the functionality of live talk and affordable price (no extra costs besides the internet connection fee). Many people with phone-talk phobia still feel comfortable to chat via Instant Messaging. The disadvantage of Instant Messaging is that there is no chance to see the client's computer screen, and it is necessary to use imagination to solve the problems and explain the results to the client.

---

[20] http://en.wikipedia.org/wiki/Instant_messaging

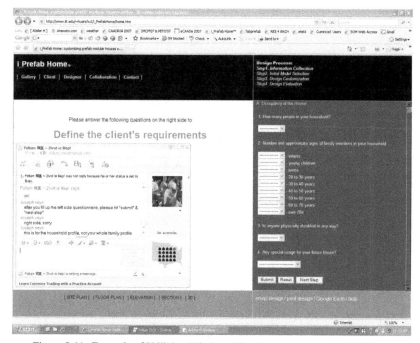

Figure 5.44. Example of Utilizing Windows Live Messenger for i_Prefab Home

5.4.3 Virtual Meeting. Virtual meeting, also called web conferencing, is used to hold group meetings or live presentations over the Internet[21]. In the early years of the Internet, the terms "web conferencing" and "computer conferencing" were often used to refer to group discussions conducted within a message board (via posted text messages), but with today's computer technology, people can not only speak/ listen via headset, but it is also possible to view the other meeting attendees via a webcam integrated to achieve a virtual meeting. Besides the personal computer's virtual meeting, some architecture firms already have a video conference room with multiple cameras, monitors, and microphones installed to host the virtual meeting for international clients and consultants. Many web conferencing

---

[21] http://en.wikipedia.org/wiki/Virtual_meeting

170

software applications provide the option to take partial control of the other meeting attendee's computer screen. For example, the "give keyboard" is an option in GoToMeeting[22] virtual meeting application. This is a more positive monitoring behavior than figuring out the problems by Instant Messaging or a telephone conversation because it can also include visual image communication.

5.4.4 User's Computer Screen Remote Controlling.   Virtual Network Computing (VNC) is a graphical user interface (GUI) desktop sharing system which uses the RFB (Remote Frame Buffer) protocol to remotely control another computer[23]. It transmits the keyboard and mouse events from one computer to another, relaying the graphical screen updates back in the other direction, over a network. A VNC system consists of a client, a server, and a communication protocol. The VNC server is the program on the machine that shares its screen. The VNC client (or viewer) is the program that watches and interacts with the server. The VNC protocol is very simple, based on one graphic primitive. By default, VNC is not a secure protocol. Besides the issue of password setting, some people might have an insecure feeling that personal information may be released by the VNC user without the right administration. It is a 100% mentoring and controlling protocol used by professional to remote help, and currently a few large size architecture firms use this technology for providing immediate IT or CAD support. Figure 5.45 is an example of using VNC technology to watch the end user's computer screen.

---

22 https://www.gotomeeting.com/
23 http://en.wikipedia.org/wiki/VNC

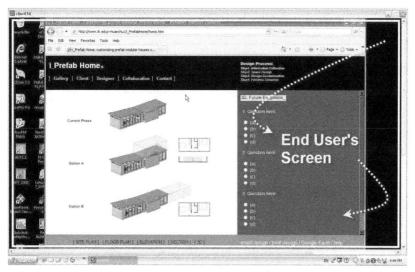

Figure 5.45. Example of Utilizing VNC to Watch the End User's Computer Screen

172

5.4.5 Design Center. A housing design center is a defined space used to present any information related to the housing product being sold. There are several different terms and formats that can represent the display theme, like showroom, reception center, or housing information center. The reason to discuss the design center here is because it is a physical space, and the optimum communication for design consulting is the face-to-face meeting. It is possible to integrate the i_Prefab advisory tool with a housing design center, and the step-by-step digital questionnaire can be accomplished by the client communication directly with a professional salesperson. In addition to the computer interface, a housing design center can also display examples of prefabricated building components, elevation materials, and interior finishes for the clients to compare and select. It can also provide tactile feelings from the real material (i.e. carpet, fabric, handle) which a computer interface could never provide. Figure 5.46 shows the photos of a housing information center in Osaka, Japan. The visitors can even use their bare feet to experience the real feeling about the flooring materials.

Figure 5.46. Photos of Housing Information Center in Osaka, Japan

The location of the housing design center is another key factor of the modular housing service. If the location is not easily accessible, it is hard to attract people and engage them in the design process. A suitable location may vary by different cultures and/or the relationship with a city's infrastructure. Most housing information centers in Japan are close to train stations (Figure 5.47), and some are located inside a housing park. The housing park in Japan is a cluster of high-quality model homes using the prefabricated construction method.

Figure 5.47. Housing Information Center near a Train Station

5.4.6 Virtual Reality. Virtual reality (VR) is computer-simulated reality. It relies on multi-sensory input and output devices such as a headset with video goggles and stereo earphones or a data glove or jumpsuit with fiber-optic sensors that tracks body movements. One can experience computer-simulated "virtual worlds" three-dimensionally through sight, sound, and touch. For example, one can enter a computer-generated virtual building to look around the spaces, and move around in this virtual environment at will.

A Computer Automatic Virtual Environment (CAVE) is an immersive virtual reality environment where projectors are directed to three, four, five or six of the walls of a room-sized cube (Figure 5.48). It can be used as a platform for a highly interactive, reactive environment that includes real-time visual feedback. This type of environment can be extended for the visualization of models as well as their construction in real time. CAVE was the first virtual reality system to let multiple users participate in the experience simultaneously. It was developed by the Electronic Visualization Laboratory at the University of Illinois at Chicago and was announced and demonstrated at the 1992 SIGGRAPH[24].

---

[24] http://en.wikipedia.org/wiki/Cave_Automatic_Virtual_Environment

Figure 5.48. Using CAVE to Simulate a Proposed Train Station Design

5.5   Building Information Modeling for Collaboration

In order to transform the client selection design information and the evaluation process by professionals (architects and engineers) with the fabrication process by the manufacturer, the virtual design geometry should be represented as an information-contained object, not just a graphic entity. Figure 5.49 demonstrates the pre-design diagnostic website (Figure 5.49, left) which is available anywhere with internet connections. All design suggestions are represented as a virtual building online and the digital design model with customized client-input data is then transformed into a Building Information Modeling (BIM) application using Autodesk Revit or a similar program (Figure 5.49, middle). After professional review, the BIM digital design model includes all of the construction

176

information and is ready for the coordination with manufacturing and the assembly of the building components in the factory. The digital design model also can be exported from the Revit application to Google Earth (Figure 5.49, right) to position directly with the exact site information and provide the client with a review of the house as a four-dimensional experience. Overall, the diagram shows the expectation of housing delivery process from web-based programming to digital design collaboration and virtual environment simulation.

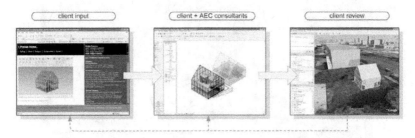

Figure 5.49. View of Prototype Interface and Work Flow Diagram

Figure 5.50 shows Revit is an object-orientated design application which is embodied with all the required information along with any necessary elements to simulate a virtual building environment before the start of the construction process. The Family Editor is an object making interface that can be used to create custom components for different parts of the building including floor panels, aluminum frames, exterior cladding, mechanical components, connectors, and so on. In many instances, custom objects are also assembled together to create larger object modules.

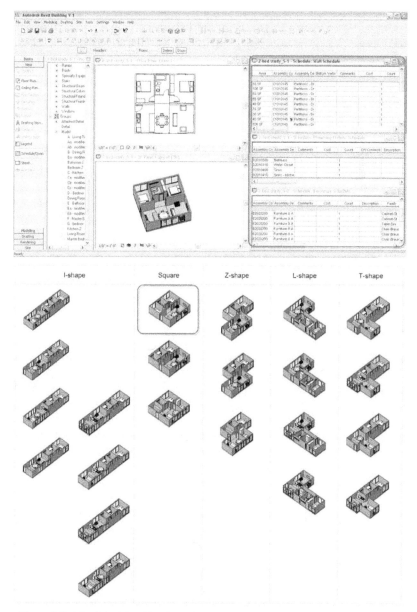

Figure 5.50. The Schedule of Modular Components is being shared with the Fabricator

Referencing KieranTimberlake's experience with Revit, it seems that BIM's object-based approach is a natural application for the modular approach to prefabricated architecture because it simplifies the process of constructing custom prefabricated buildings. The objective is to overcome the "cookie-cutter" uniformity and monotony which have given prefabrication a bad reputation in the past (Khemlani, 2005).

e-SPECS is software to automate the preparation of construction specifications for architects. There is a direct link between e-SPECS and the Revit database that generates more cost savings and helps guarantee that the information in the project specifications match the current state of the building model.

5.5.1 Digital Information Models from Suppliers. Unlike other large industries of aerospace or the automobile industry, the building industry is highly fragmented with most projects undertaken by one time only project-based organizations consisting of the many different individual roles of client, architect, engineer, contractor, and suppliers. There is a high risk of human errors generated during the communication and exchanging of data among these individual groups. Ironically, the architects may use computer-aided design to produce the drawings required for bidding and construction, but these digital work products are not necessarily useful for the tasks performed by the contractor: costing and scheduling. This problem is caused because computers have been applied to automating specific tasks rather than addressing the overall building process (Cohen, 2005).

Networked computing presents the opportunity to integrate information from many sources during the project delivery process. If most of the individual prefabricated modular housing components could be standardized for interchanging and manufacturers or suppliers could provide the information of modular building elements as an interchangeable digital format on their website to represent their products as accurately as possible for the final construction assembly, modular housing systems could then be developed by assembling

179

these components to create many different compositions. A new design configuration could be created virtually during design as well as physically during offsite fabrication and onsite installation to achieve the optimistic goal of mass customization of prefabrication without sacrificing time efficiency and would be competitive the cost affordability that already exists in mass produced housing industry.

Here are two examples to restate this proposed concept. Currently, Autodesk Revit Architecture 2008 Content Distribution Center provides many families for users to download (Figure 5.51). These families can be viewed as parametric information objects as well as three-dimensional geometrical representations. However, the usage of these intelligent "blocks" is still being treated as traditional blocks in most architecture firms. Since Autodesk provides the information as generic references from its library, architects still need to customize any selected Revit family to become a real constructible object based on the supplier's shop drawing. Some mistakes and time consuming may exist during this model reproducing process.

Figure 5.51. The Families of Vanity Provides by Autodesk Revit Library

To avoid many human errors and time consuming during the design development process, the product suppliers or vendors should provide 3D information imbedded digital models from their website, as a supplemental to the shop drawings. Herman Miller, Inc., a global provider and manufacturer of office furniture and equipment, provides 3D digital models from the product showroom of their website (Figure 5.52). Moreover, the website utilizes i-drop technology to provide the drag-and-drop capability for the users to apply their products easily to the digital design platform, like Autodesk 3D Studio Max, AutoCAD, or Revit (Figure 5.53).

i-drop is an Autodesk's invention for publishing design objects within standard Web pages. Web pages that contain i-drop objects look and behave exactly like a standard Web page in a standard Web browser. When Web pages are displayed in any of Autodesk's Web-enabled desktop design tools, an i-drop object becomes an open gateway between the content provider's server and the Autodesk user's desktop. This promises to vastly accelerate the process of manufacturers used to conduct business with those that are responsible for evaluating and specifying their products (Autodesk, 2003). In other words, i-drop is a mechanism to enable Autodesk digital design applications users to use the intelligent blocks directly from the provider's websites. Instead of having to search for an object on the web, download the digital files, and then inserting that into the platform of digital design applications, this mechanism simplifies the process for the convenience of the users.

Figure 5.52. Integration of i-drop from Herman Miller's Website

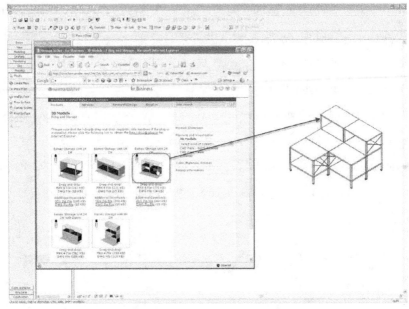

Figure 5.53. The Capability of "drag-and-drop" from Web page to Revit

183

5.5.2 Digital-based Housing Delivery System.   Although the majority of the current architectural design process already has been shifted from the traditional paper-based method to an advanced digital-based method, the problem of delay and error output in the project delivery process can still exist. The building process would be much better served if the entire chain of information from design to construction to operations could remain in one seamless digital format (Cohen, 2005). Using the expanded i_Prefab theoretical model as an example, a client inputs spatial requirements, an advisory tool organizes the client's need and generates available design suggestions, an architect reviews and revises, an engineer calculates, and a construction/ fabrication manager schedules, all using information from a common project database that is accessible over a network (Figure 5.54).

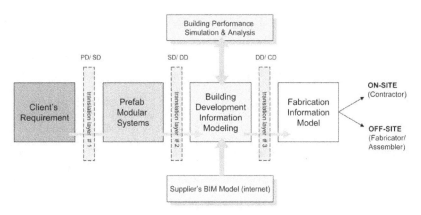

Figure 5.54. Value-Enhanced BIM in Modular Housing Delivery Process

5.6   Virtual On-site Design Review

In the last section, Chapter 5.8 introduced the digital design model that was generated from the i_Prefab advisory tool has the option to transfer to a BIM application like Autodesk Revit for further modifications. Moreover, the professional modified digital model also can be exported from Revit platform to Google Earth for the clients to preview their future house with the virtual building site context from anywhere of the world with an internet connection.

There are three reasons for integrating Google Earth as an alternative for design review:

1. Virtual on-site simulation for the clients and professional participants to have a four-dimensional experience (i.e. fly-through, walk-in) to review the design before the manufacturing process.

2. Consideration of not only the individual house design, but also the harmony of integrating the new house into the neighborhood.

3. Contributing and constructing the up-to-date virtual design buildings to the global database for future reference (public and private use).

In order to post the digital design model from Revit to Google Earth, professional users need to download a plug-in, called Avatech Earth Connector[25], an Add-In for Google Earth. The software created by Avatech Solutions is currently only in the BETA version. The plug-in provides a user-friendly wizard inside of the Revit toolbar to go through the simple process of file conversion from RVT (Revit file format) to KMZ (Google Earth file format) (Figure 5.55). The most important issues to correctly show the Revit model exactly to the virtual building site in Google Earth are location and orientation. The accurate data of longitude and latitude can be easily acquired by adding a new Placemark in any virtual location of Google Earth. Every Placemark in Google Earth can be saved as KML-type file. The file format can be imported to Avatech Earth Connector for Revit even without remembering any numbers.

---

[25] http://www.avat.com/products/software/avatech/earthconnector/

Figure 5.55. Avatech Earth Connector – Revit to Google Earth Wizard

Figure 5.56 is an example of how to bring Revit Model to Google Earth for Design Review. Since the plug-in connector contains all BIM data when exporting from Revit to Google Earth, professional users can easily control the layers of each individual building component (i.e. roof, wall, curtain wall panel, floor, stair) to see the interior space and do further modifications. The design review process may then go back and forth between Revit and Google Earth.

Figure 5.56. Example of Revit Model to Google Earth for Design Review

Besides the design review by a professional, the clients can access Google Earth to see their future house design in two ways:

1. The architect can send the KMZ file (Zipped Keyhole Markup Language for Google Earth) directly to the client via email. Once the client receives the email, he or she

187

can open the attached KMZ file directly in any computer with Google Earth installed and internet connections.

2. The architect can create a unique name as the project ID and post the digital design model on Google 3D Warehouse. The client can be given the project ID and go to search the house from Google 3D Warehouse website[26] (Figure 5.57).

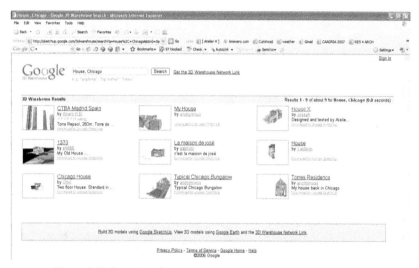

Figure 5.57. Example of Searching "House" in Google 3D Warehouse

When the client finds the result of the proposed digital design model, the webpage of Google 3D Warehouse looks like Figure 5.58. The client can view the model with virtual site context by clicking the "View in Google Earth 4" button. The Google Earth application will bring the viewer from macro scale of the earth to micro scale of the city (Figure 5.59) and the specific site location with digital model integrated can be produced in just ten seconds (Figure 5.60).

---

[26] Google 3D Warehouse - http://sketchup.google.com/3dwarehouse/

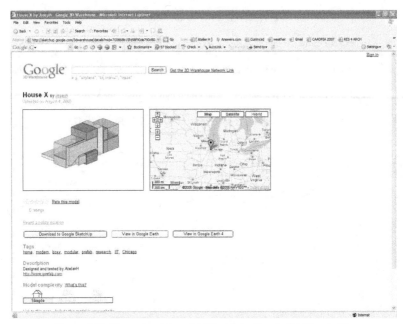

Figure 5.58. Example of Searched Result in Google 3D Warehouse

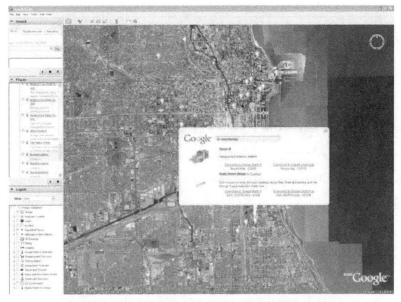

Figure 5.59. Google Earth with 3D Warehouse Network Link

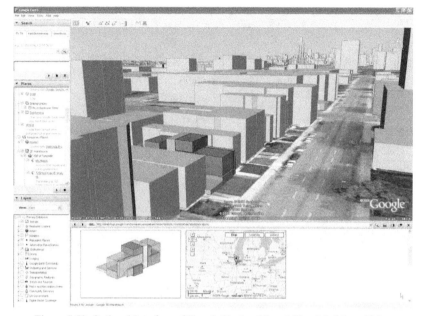

Figure 5.60. Optional Interface of Google Earth – Virtual City, Model, and Map

The 3D Warehouse is like a global virtual building database, and keeps growing day by day. If the client has a basic knowledge level of 3D computer modeling, he or she even can also download the architect's designed model to Google SketchUp (Figure 5.61) for reference or as a platform to discuss with professional consultants.

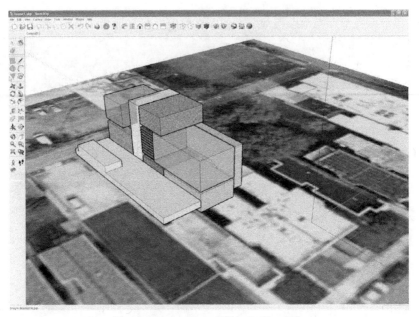

Figure 5.61. Interface of Google SketchUp

5.7   Output Formats and Data Delivery

The definition of output format here is to describe the final presentation of i_Prefab Home design results to the client. It also covers the different formats of digital data to be exchanged between AEC consultants and acceptable file formats for digital fabrication. Figure 5.62 shows three different kind of data during the entire design process: (1) PDF documentation at the end of the web-based schematic design phase for the clients, (2) Building Information Modeling (BIM) data for coordination and analysis between AEC consultants, and (3) Computer-Aided Manufacturer (CAM) data for digital fabrication and a list of quantities for the manufacturer.

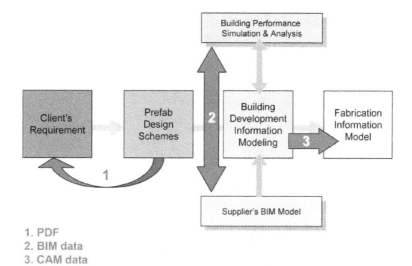

1. PDF
2. BIM data
3. CAM data

Figure 5.62. Output Formats and Data Delivery in the Design Process

5.7.1 Output Data for the Client. At the lower-right corner of the i_Prefab Home web interface, there is a button called "print design". After clients finish the schematic design process online, the end results can be generated as a Portable Document Format (PDF) that allows printing and emailing to someone easily. The format is invented by Adobe Systems over 15 years, and as a very common light-weight digital format to transfer the data with text and images. This document may include the design plans, configuration list, 3D views, and contract (Figure 5.63).

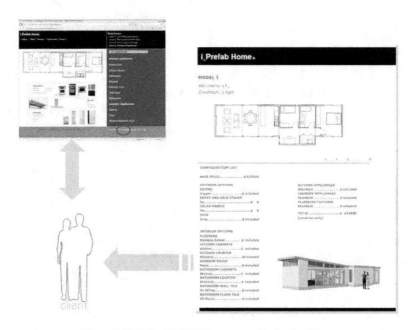

Figure 5.63. Printable PDF Documentation for the Client

5.7.2 Design Coordination between AEC Consultants. One of the benefits of using Building Information Modeling (BIM) as the digital design and documentation platform is the inter-disciplinary design coordination. Since all design elements are 3-dimenional objects from architects, structural engineers, and MEP consultants, it can be more accurate to check

193

the interferences in a 3D environment. Figure 5.64 shows the design coordination and clash detection in Autodesk NavisWorks software. The application can unite all design contributions into a single, synchronized building information model as the heart of BIM workflow.

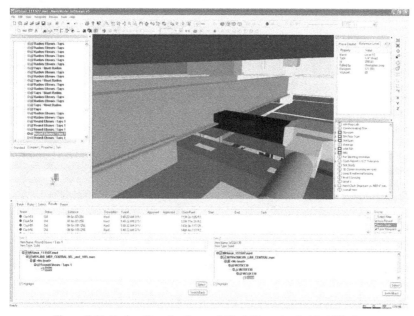

Figure 5.64. Design Coordination and Clash Detection in NavisWorks

5.7.3 Output Data for the Manufacturer. Generally speaking, there are two different types of output data for the manufacturer, one is the list of quantities, and the other one is the Computer-Aided Manufacturer (CAM) data for digital fabrication. Unlike the traditional architectural design, the construction documentation phase can be simplified by rearranging the prefabricated modular components from the factory. Figure 5.65 illustrates the BIM application – Autodesk Revit can generate a list of quantities for cost estimate and inventory of modular components before the manufacturing process. Any custom-made free forms can

194

be exported directly and constructed by a CNC machine with a plug-in inside of the digital design tool like Rhinoceros 3D.

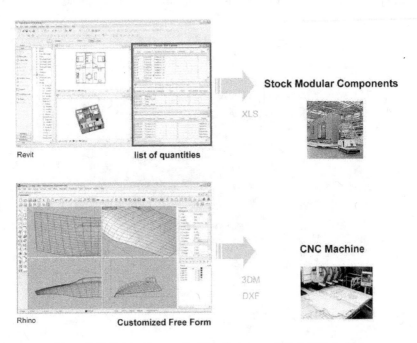

Figure 5.65. Output Formats for Manufacturer and Digital Fabrication

CHAPTER 6

CONCLUSIONS

This thesis proposes a new methodology of integrating client's requirements and prefabricated building components from selected available modular housing systems driven by web-based questionnaire. The proposed interface and system work flow support a dynamic experience for consumer participatory design through digital interaction. It is believed that the findings of this research provide a hierarchy of questions resulting in the spatial conversion from questionnaires to architectural simulation to non-professional end users. It will serve as a valuable reference for future research and software development in the area of mass-customized, consumer-involved modular housing industry. This final chapter will evaluate the research presented in this thesis, including its achievements, limitations, and suggest possible directions for future research.

6.1   Achievements

The goal of present research was to investigate the possibilities of applying the mass customization concept into the prefabricated modular housing industry. Although the concept has already been implemented in the computer, clothing, and automobile industries, it has not been fully integrated in architecture, especially in the housing industry which is more directly related to personal life style. One of the arguments in this thesis focused on the current implementations in the other industries that are not fully functional to address the consumer's needs by selecting the product's appearance only. The research presents the possibilities of customizing modular houses by internet and architectural programming process beyond merely a catalog selection from the existing prefabricated housing vendors.

The comparisons of the original research objectives and achievements are concluded as below:

1. Objective: To research how to collect and interpret client's need to formulate design options via internet in order to address the issues of individual needs from the consumers.

Result: Defined the methods of collecting client's requirement by digital questionnaire input and explored the content of digital questionnaire regarding the available design configurations of selected modular systems. Tested the Personal Brain application as a decision tree study sketch platform to combine the client's need and the modular house design configurations (Refer to Chapter 4.4 and Chapter 4.7-4.8).

2. Objective: To explore possible combinations of prefabricated modular house according to client's preference.

Result: Analyzed existing prefabricated modular house systems and explored possible configurations in two case studies of two-bedroom house and four-bedroom house. Interpreted the client's need to adopt, modify, and recreate different configurations from the available models of the selected vendors; revised the modular unit to be more flexible and client responsive (Refer to Chapter 4.6).

3. Objective: To suggest a framework of mass-customized prefabricated housing delivery system and simulate the process from the definition of client's preference to the ready-to-build housing components.

Result: Illustrated the system work flow of the proposed model, which includes basic mode for the client and professional mode interaction for the design consultants, and the entire process from the pre-design to the construction phase. Forecasted Building Information Modeling (BIM) application as a design collaboration platform among architects, engineers, and manufacturers in the building delivery process, and this especially has a potential benefit for the modular housing industry (Refer to Chapter 4.5 and Chapter 5.5).

4. Objective: To propose a method of web-based collaboration for the client, architect, and manufacturer to integrate the process of mass-customized prefabricated house online.

Result: The i_Prefab advisory system has been created as a web-based design consulting tool and provides the capability for the client to customize the future house online. This prototype brings prefabricated modular housing to the next level of mass customization with digital technology and bridges the gap between non-professional (client) and professional (architect) by representing architectural knowledge via web media (Refer to Chapter 5).

Overall, the research combines the theoretical idea of mass customization with the practical application of digital design and prefabrication technology. Addressing a particular problem in the design of prefabricated modular housing allows us to see bigger issues in participatory design approaches. The results also demonstrate the power of internet to communicate between design teams and integrate products and materials.

6.2   Limitations

The limitations of this research are twofold: one is the component exchangeability between different building systems, and the other is the content of questionnaire in the web-based prototype. First, the case studies of design configurations in this research are all focused on the same modular system of the selected prefabricated housing vendor. It would be hard to examine the possibilities of combining modular components from multiple vendors, though theoretically it is possible with by providing a uniform general connector between different modular housing products. However, such connectors are not yet ready in the housing industry today, especially in the free market countries like the United States. The general connector idea for modular building components would be similar with the computer industry's USB port and which supports or integrates modular components to plug-and-play.

198

Second, the content of digital questionnaire itself has some constraints related with local language, culture, and people's preconceptions. The concept of using a questionnaire to collect the client's requirements is a method that can be customized by the local vendors. Since architecture is related with place and people, the questionnaire should reflect the local climates and site context, plus the culture of living style. It is inappropriate to attempt to create a single universal questionnaire to put on the World-Wide Web for different people even though the language is not a barrier. Besides, this research presents some difficulties of eliciting accurate information and telling people's mind unless the questions are rich enough to cover everything and the design systems are flexible enough to meet the different clients. However, the research unveils the follow-up research problems regarding the psychology of learning and human-machine interface.

6.3   Future Research

The results of the thesis research establish an open-ended framework of a design consulting and decision support system by a dynamic web-based questionnaire methodology for future researchers. The achieved research objectives bring prefabricated modular housing design to the next level of integrating the goal of mass customization with digital technology. The interpretation of translating the client's need to match different spatial configurations of the design models from the selected prefabricated modular housing vendors is based on the architect's assumption. However, this is an online recommendation to replace the limited face-to-face first meeting time between architects and clients. Revising the modular unit to be more flexible and client responsive is the recommendation to the modular housing design architects and vendors. It is obvious that costs can be reduced in the architectural design and engineering consulting phases. Moreover, the factory-made modular houses can save the material waste and provide a better quality and time control rather than the traditional construction method with its tremendous labor cost. There are several possible directions in which the research can proceed to explore and further refine the concept of enabled seamless design collaboration in the housing delivery process. The following topics are the potential approaches:

1. Rule-based design configuration for multiple stories of prefabricated single-family house. The considerations of vertical circulation and building code restrictions will be integrated with programming in order to create a fully consumer participatory space planning tool for advance users.

2. To investigate the output format for manufacturing from digital design applications, like Rhino or Digital Project, and develop the bi-directional linkage between the i_Prefab advisory system and Building Information Modeling applications to make as a web-based feedback loop between non-professional users and professional users.

3. Further development of the interactive questionnaire approach coupled with other BIM related programming software (i.e. ONUMA Planning System) to establish improved communication between consumers and design professionals in all types of buildings.

4. To investigate the potential technology of Virtual Reality (VR) and the economical format of VR environment to be integrated as an architectural design-aided tool.

5. From digital design to digital fabrication for modular house development: to explore and improve the current limitations of digital fabrication, like the facility cost, output size limitation, transportation cost and limitation, labor union against, building system integration, etc.

6. To investigate the potential prefabricated materials for digital fabrication like SIP (Structural Insulated Panels) with the concerns of customizability and availability.

7. Prefabricated modular housing design toward sustainable development. Strategies like using building-integrated photovoltaics (BIPV), sustainable and durable materials, and renewable "green" power (i.e. wind) will add more value to the result of current research.

8. To elaborate the consumer participatory concept model for multi-family housing and integrate Open Building approach of implementing the infill-support system (Refer to Chapter 4.9). This mass customization concept can also be applied to design a high-quality affordable housing or emergency housing after disaster for those people who usually do not have a chance to be entitled to deserve a house based on their own preferred input.

6.4   Closing Remarks

Developing knowledge by means of doing research is endless. The goal of this research is to propose a consumer participatory procedure – dynamic questionnaire to convert the client's needs and preferences to become buildable housing design options. A problem-solving approach has been accomplished in the methodology development, analysis, and case studies so that a complete framework of digital-based prefab housing design delivery workflow has been established for the current prototype and future research. A web-based design advisory tool cannot totally replace a human architect in the schematic design phase when the client has no idea or has too many ideas beyond the limitation of predetermined data. This research does not attempt to change the traditional role of architect but rather to extend the architecture knowledge into digital and public-accessible environments. Even so, the challenges have been tremendous due to technical, social, and professional obstacles to adopt this digitally mediated communication method. No matter the query approach method can or cannot intelligently solve the design customization issue, the research documentation can be contributed as a more user-oriented guideline for the prefabricated house design with today's technology.

APPENDIX A

QUESTIONNAIRE V.1 AND SPATIAL CONVERSION

[Questionnaire of i_Prefab Home Design]

Family Profile

Question	Answer	Spatial Conversion
1. How many people in your household?	___ 1 ___ 2 ___ 3 ___ 4 ___ 5 ___ 6 or more	N/A (basic data)
2. Number and approximate ages of family members in your household:	___ Infants ___ Young children ___ Teens ___ 20 to 30 yrs ___ 31 to 40 yrs ___ 41 to 50 yrs ___ 51 to 60 yrs ___ 61 to 70 yrs ___ 70+	N/A (basic data)
3. What will be the best way to describe your household configuration?	___ Single ___ Couple (married or unmarried) ___ Parents + 1 child ___ Parents + 2 children ___ Parents + 3 children ___ Single parent + 1 child ___ Single parent + 2 children ___ Single parent + 3 children ___ Parents + child + grandparent ___ Single parent + 2 children + grandparent ___ Other___	1 bedroom 1 large bedroom 2 bedrooms (1 large + 1 small) 3 bedrooms (1 large + 2 small) 4 bedrooms (1 large + 3 small) 2 bedrooms (2 small) 3 bedrooms (3 small) 4 bedrooms (4 small) 3 bedrooms (1 large + 2 small) 4 bedrooms (4 small)

4. Is anyone physically disabled in any way?	__ Yes __ No	1 story suggested option, ADA bathroom
5. Are you planning to add a new member in next three years?	__ Yes __ No __ Not sure	Add 1 bedroom

Lifestyle

Question	Answer	Spatial Conversion
6. Do you regularly prepare meals in your kitchen rather than eat out?	__ Most of time __ Sometimes __ Rarely	More countertop space Standard (Default setting) Less countertop space
7. Where does your family eat its meals?	__ Kitchen __ Dining Room/ Dining Area __ Family Room __ Other	Eating area required Adjacent to kitchen Adjacent to kitchen
8. Do you need a formal dining room for special occasions?	__ Yes __ No	Dining room required Dining area as part of family room

204

Question	Answer	Spatial Conversion
9. What is the primary cooking style in your family?	___ Family Meals ___ Quick & Simple Meals ___ Bringing Meals Home (microwave only)	Large size kitchen Medium size kitchen Small size kitchen
10. What does the primary cook prefer?	___ No one else in the kitchen while preparing meals ___ A helper in the kitchen when preparing meals ___ Family or friends visiting during meal preparation	Gallery or single bar kitchen layout L-shape kitchen layout L-shape w/ island kitchen layout
11. How much food do you buy at once?	___ A lot (for the week) ___ A little (for each meal/ day)	Extra cabinets & large refrigerator
12. Do you need a multi-media room for entertainment	___ Yes ___ No	Multi-media room required
13. Where do you work?	___ Always at the office ___ Sometimes at home ___ Always at home	Den or small library Home office
14. How do you feel about guest in your home?	___ Plan to have a guest room ___ Plan to have a folding bed in multi-functional room (home office) ___ Welcome, but do not prefer to stay over night	Guest room Folding bed in somewhere

15. Where do you plan to park your vehicles?	___ On the street ___ Carport ___ Attached Garage ___ Detached Garage	No garage Carport required Attached Garage required Detached Garage required
16. How many vehicles do you plan to park in Garage/ Carport (based on question 15)?	___ 1 ___ 2 ___ 3	

Site Context

Question	Answer	Spatial Conversion
17. Do you have a site purchased?	___ Yes ___ No	Location
18. If yes, what is the context of the site?	___ Urban infill ___ Suburban ___ Rural	Lot size constraint, zoning code Lot size constraint, zoning code
19. What is the character of the site?	___ Flat ___ Sloping ___ Irregular ___ Other	Relationship of ground level and land More definitions
20. How important are the surrounding views to you?	___ Not really ___ Very ___ Most	Minimize opening/ skylight/ courtyard

21. Where are the best views of the site?	___ North ___ Northeast ___ East ___ Southeast ___ South ___ Southwest ___ West ___ Northwest	(opening orientation/ yard location/ master bedroom location)
22. If this is an infill site, the front of the house (adjacent to the street) will face:	___ North ___ East ___ South ___ West	Sun shading required

Building Geometry

Question	Answer	Spatial Conversion
23. The perimeter plan geometry should be:	___ Regular geometry with minimum wall projections ___ Medium Irregular plan with meandering exterior walls ___ Complex	(find matched housing types from catalog)

Construction Budget

Question	Answer	Spatial Conversion
24. What is your total estimated construction budget of new home (not include any land cost)?	___ Under $200,000 (USD) ___ $200,000 - $300,000 (USD) ___ $300,000 - $400,000 (USD) ___ $400,000 - $500,000 (USD) ___ $500,000 - $600,000 (USD) ___ $600,000 - $700,000 (USD) ___ $700,000 + (USD)	(find matched price range from catalog as a starting plan)

APPENDIX B

QUESTIONNAIRE V.2

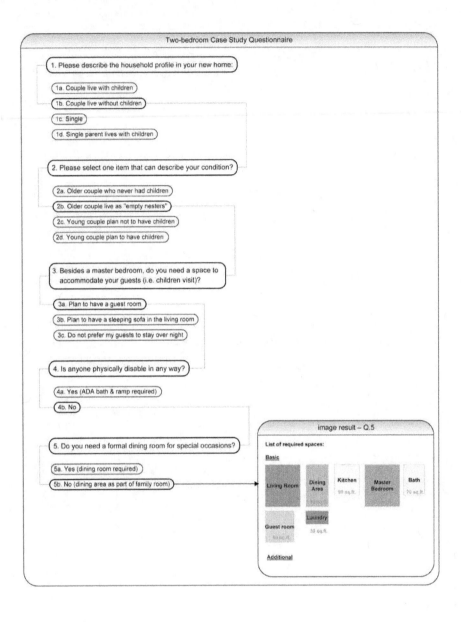

Two-bedroom Case Study Questionnaire

1. Please describe the household profile in your new home:

1a. Couple live with children

1b. Couple live without children

1c. Single

1d. Single parent lives with children

2. Please select one item that can describe your condition?

2a. Older couple who never had children

2b. Older couple live as "empty nesters"

2c. Young couple plan not to have children

2d. Young couple plan to have children

3. Besides a master bedroom, do you need a space to accommodate your guests (i.e. children visit)?

3a. Plan to have a guest room

3b. Plan to have a sleeping sofa in the living room

3c. Do not prefer my guests to stay over night

4. Is anyone physically disable in any way?

4a. Yes (ADA bath & ramp required)

4b. No

5. Do you need a formal dining room for special occasions?

5a. Yes (dining room required)

5b. No (dining area as part of family room)

image result – Q.5

List of required spaces:

Basic

Living Room

Dining Area

Kitchen
98 sq.ft.

Master Bedroom

Bath
70 sq.ft.

Guest room
55 sq.ft.

Laundry
35 sq.ft.

Additional

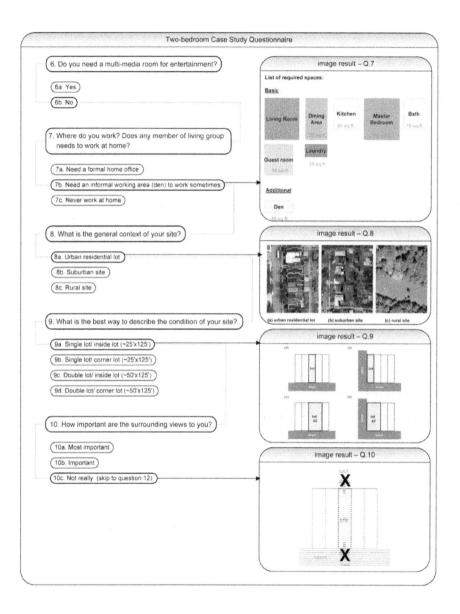

Two-bedroom Case Study Questionnaire

6. Do you need a multi-media room for entertainment?

6a. Yes

6b. No

7. Where do you work? Does any member of living group needs to work at home?

7a. Need a formal home office

7b. Need an informal working area (den) to work sometimes

7c. Never work at home

8. What is the general context of your site?

8a. Urban residential lot

8b. Suburban site

8c. Rural site

9. What is the best way to describe the condition of your site?

9a. Single lot/ inside lot (~25'x125')

9b. Single lot/ corner lot (~25'x125')

9c. Double lot/ inside lot (~50'x125')

9d. Double lot/ corner lot (~50'x125')

10. How important are the surrounding views to you?

10a. Most important

10b. Important

10c. Not really (skip to question 12)

image result – Q.7

List of required spaces:

Basic

Living Room | Dining Area | Kitchen | Master Bedroom | Bath

Guest room | Laundry

Additional

Den

image result – Q.8

(a) urban residential lot   (b) suburban site   (c) rural site

image result – Q.9

image result – Q.10

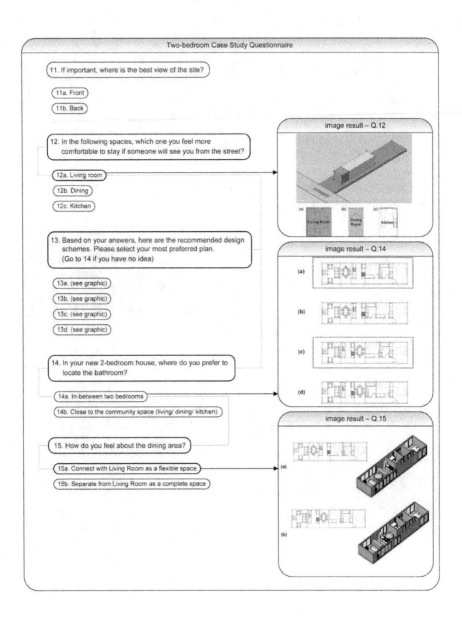

Two-bedroom Case Study Questionnaire

11. If important, where is the best view of the site?

11a. Front
11b. Back

image result – Q.12

12. In the following spaces, which one you feel more comfortable to stay if someone will see you from the street?

12a. Living room
12b. Dining
12c. Kitchen

13. Based on your answers, here are the recommended design schemes. Please select your most preferred plan.
(Go to 14 if you have no idea)

13a. (see graphic)
13b. (see graphic)
13c. (see graphic)
13d. (see graphic)

image result – Q.14

(a)
(b)
(c)
(d)

14. In your new 2-bedroom house, where do you prefer to locate the bathroom?

14a. In-between two bedrooms
14b. Close to the community space (living/ dining/ kitchen)

15. How do you feel about the dining area?

15a. Connect with Living Room as a flexible space
15b. Separate from Living Room as a complete space

image result – Q.15

(a)
(b)

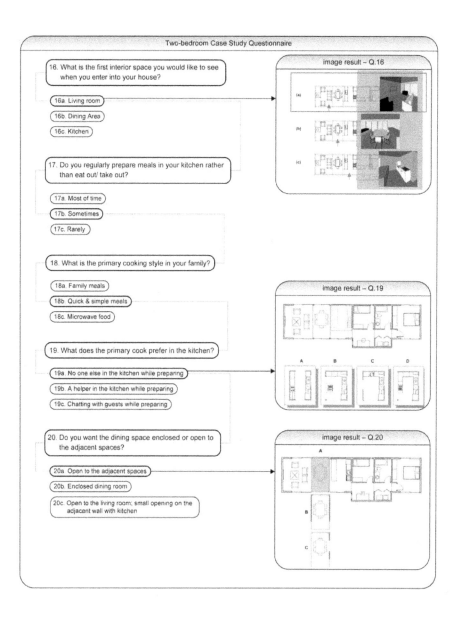

Two-bedroom Case Study Questionnaire

16. What is the first interior space you would like to see when you enter into your house?

16a. Living room
16b. Dining Area
16c. Kitchen

image result – Q.16

(a)
(b)
(c)

17. Do you regularly prepare meals in your kitchen rather than eat out/ take out?

17a. Most of time
17b. Sometimes
17c. Rarely

18. What is the primary cooking style in your family?

18a. Family meals
18b. Quick & simple meals
18c. Microwave food

image result – Q.19

19. What does the primary cook prefer in the kitchen?

19a. No one else in the kitchen while preparing
19b. A helper in the kitchen while preparing
19c. Chatting with guests while preparing

A   B   C   D

20. Do you want the dining space enclosed or open to the adjacent spaces?

20a. Open to the adjacent spaces
20b. Enclosed dining room
20c. Open to the living room; small opening on the adjacent wall with kitchen

image result – Q.20

A
B
C

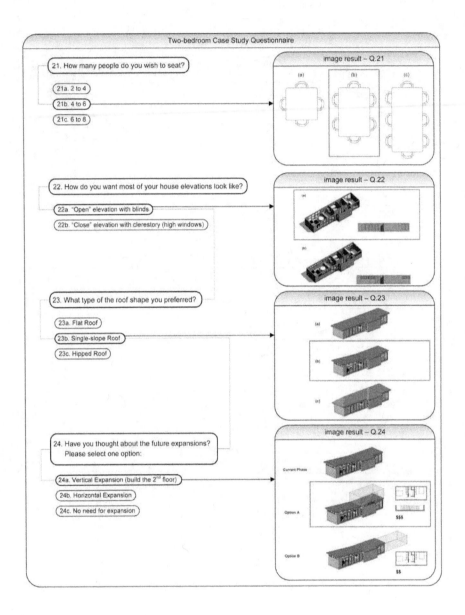

Two-bedroom Case Study Questionnaire

image result – Q.21

21. How many people do you wish to seat?

21a. 2 to 4
21b. 4 to 6
21c. 6 to 8

(a) (b) (c)

image result – Q.22

22. How do you want most of your house elevations look like?

22a. "Open" elevation with blinds
22b. "Close" elevation with clerestory (high windows)

image result – Q.23

23. What type of the roof shape you preferred?

23a. Flat Roof
23b. Single-slope Roof
23c. Hipped Roof

image result – Q.24

24. Have you thought about the future expansions? Please select one option:

24a. Vertical Expansion (build the 2nd floor)
24b. Horizontal Expansion
24c. No need for expansion

Current Phase

Option A $$$

Option B $$

213

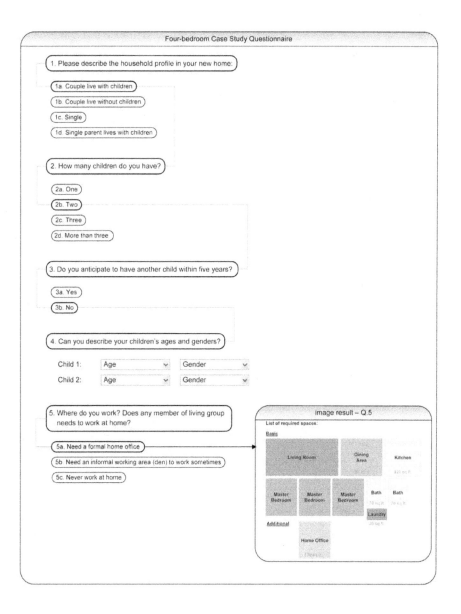

6. Do you need a formal dining room for special occasions?

6a. Yes (dining room required)

6b. No (dining area as part of family room)

7. Do you need a multi-media room for entertainment?

7a. Yes

7b. No

8. How do you want to accommodate your guests?

8a. Plan to have a guest room

8b. Plan to have a sleeping sofa in the living room

8c. Do not prefer my guests to stay over night

9. Based on the previous question, can the home office double as a guest room?

9a. Yes

9b. No

10. Will you need a basement for laundry room, storage, and playroom?

10a. Yes

10b. No

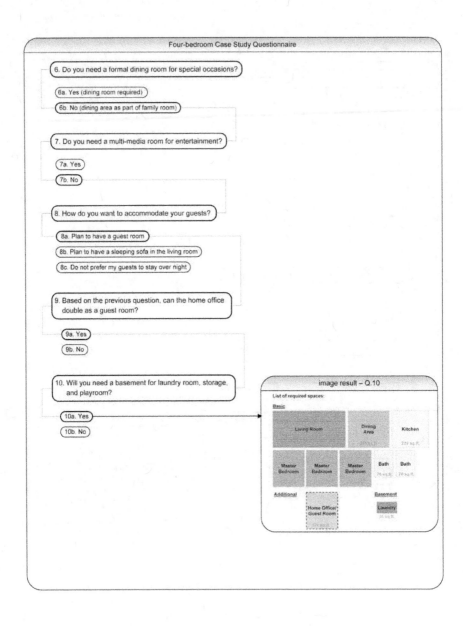

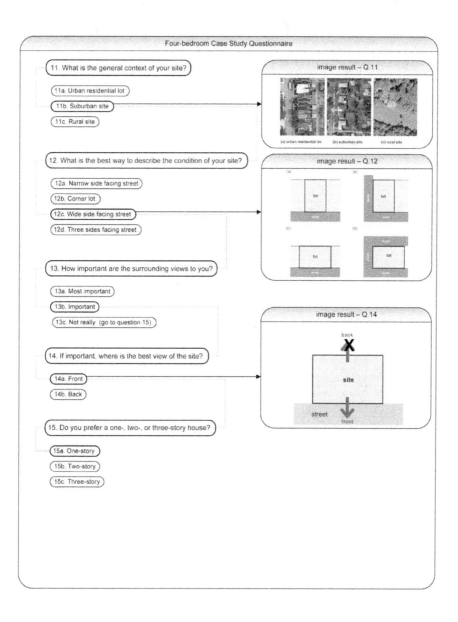

Four-bedroom Case Study Questionnaire

11. What is the general context of your site?

(11a. Urban residential lot)

(11b. Suburban site)

(11c. Rural site)

image result – Q.11

(a) urban residential lot   (b) suburban site   (c) rural site

12. What is the best way to describe the condition of your site?

(12a. Narrow side facing street)

(12b. Corner lot)

(12c. Wide side facing street)

(12d. Three sides facing street)

image result – Q.12

13. How important are the surrounding views to you?

(13a. Most important)

(13b. Important)

(13c. Not really (go to question 15))

14. If important, where is the best view of the site?

(14a. Front)

(14b. Back)

image result – Q.14

back
X
site
street
front

15. Do you prefer a one-, two-, or three-story house?

(15a. One-story)

(15b. Two-story)

(15c. Three-story)

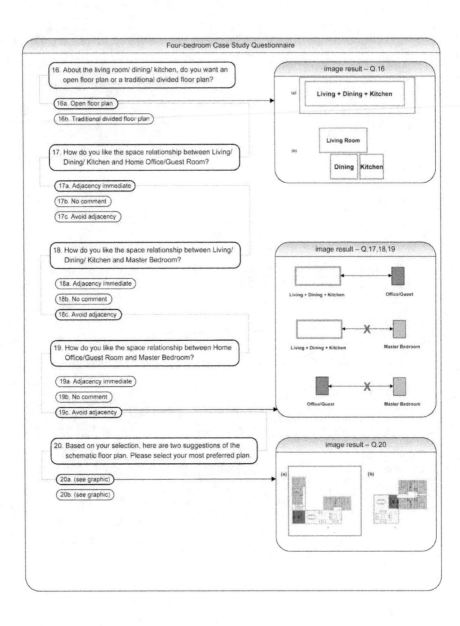

Four-bedroom Case Study Questionnaire

16. About the living room/ dining/ kitchen, do you want an open floor plan or a traditional divided floor plan?

16a. Open floor plan
16b. Traditional divided floor plan

image result – Q.16

(a) Living + Dining + Kitchen

(b) Living Room

Dining | Kitchen

17. How do you like the space relationship between Living/ Dining/ Kitchen and Home Office/Guest Room?

17a. Adjacency immediate
17b. No comment
17c. Avoid adjacency

18. How do you like the space relationship between Living/ Dining/ Kitchen and Master Bedroom?

18a. Adjacency immediate
18b. No comment
18c. Avoid adjacency

19. How do you like the space relationship between Home Office/Guest Room and Master Bedroom?

19a. Adjacency immediate
19b. No comment
19c. Avoid adjacency

image result – Q.17,18,19

Living + Dining + Kitchen   Office/Guest

Living + Dining + Kitchen   Master Bedroom

Office/Guest   Master Bedroom

20. Based on your selection, here are two suggestions of the schematic floor plan. Please select your most preferred plan.

20a. (see graphic)
20b. (see graphic)

image result – Q.20

(a)   (b)

21. Would you prefer an attached or detached garage?

21a. Attached garage

21b. Detached garage

22. How many cars do you and your family members drive regularly?

22a. Two

22b. Three

22b. Four

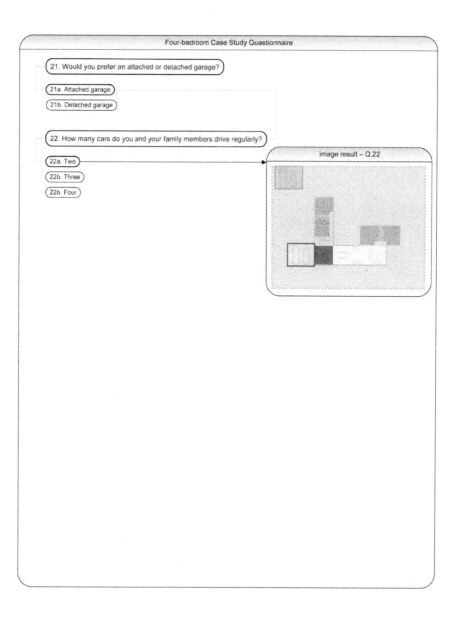

image result – Q.22

APPENDIX C

PROTOTYPE SURVEY AND USER FEEDBACK

Survey Result of Prototype User Tests

The following questionnaires are the prototype survey of i_Prefab Home design advisory tool. The purpose of this survey is to get the user feedback for validating the proposed concept and revising the web-based design interface. The four questions in the survey cover user empirical feedback on the prototype of finding suitable modular house design solutions based on the customer's input. The results may improve the interface to be more user-friendly and further concerns of the interface design strategies for the future research.

1. From your use experience, how different do you think between typical face-to-face design consulting with an architect and the web-based home design approach like this?

2. Do you think the method of using dynamic questionnaire to collect client's requirement and providing the matched design solutions online is possible and successful?

3. Do you consider the i_Prefab Home (interface) navigation process is intuitive? What can be improved?

4. As a consumer, if you have a chance to participate in your home design process from the beginning, what is the most important issue from the following options and why? (1) List of the required spaces; (2) Plan layout; (3) Individual Room Design; (4) Finishes and Appliances.

1. From your use experience, how different do you think between typical face-to-face design consulting with an architect and the web-based home design approach like this?

    I believe the uncertainty and surprising is the character of face to face design consulting, compare to a pre-defined web-based home. Traditional architect can discuss with clients, share the ideas, architects may also have more power to influence clients, and clients may state their vision more clear. Web-based is more like a selection and matrix of combinations. It is fun. The challenge for this kind of design process is designing elements and composition of modular. For factory, they should be unified, easy assembling; for architects, they should be flexible. As you mentioned, only 5% people hire an architect, people who may have interests in your approach must have some money and want to do something different from the rest 95%. So my question is your client target. For example, if the old couple they don't know how to use internet well, they don't understand the drawings or perspectives, can they trust this website to decide their future homes?

2. Do you think the method of using dynamic questionnaire to collect client's requirement and providing the matched design solutions online is possible and successful?

    Yes, I like this concept. Individualization is widely used in industry, website communication, fashion design and so on. Many websites use personal questionnaire to profile their custom in order to serve them more personalized way. Examples can be seen from I-Google. I believe the questionnaire is a good tool to collect client's requirement including their life style, their interests, and their requests.

3. Do you consider the i_Prefab Home (interface) navigation process is intuitive? What can be improved?

    I like the interface, the procedure is clear, the selection is detail. However, there several things can be improved. 1) In the design process, there is no <back> button, which I can't go back to see my selection in case I forget or I want to change my decision. 2) There is no element that may change during time. For example, the young adults have two kids right now, they need three bedroom. After 4 years, one kid grows up, he moves out, his room left empty. "How does the family change his bedroom into other functions" is not suggested in the website. 3) The cost is not clearly stated in the website. Cost is one of the important factors that influence people's decision. 4) There is no clear link between the gallery section to client section. How do I contact architect if I want something in the gallery?

4. As a consumer, if you have a chance to participate in your home design process from the beginning, what is the most important issue from the following options and why? (1) List of the required spaces; (2) Plan layout; (3) Individual Room Design; (4) Finishes and Appliances.

    Plan layout. This is most important for me, and I need to consider the layout for each room. What's their views, that's their spatial characters, what's the flows from one room to another, what's the perspective from individual point, what's private and public space, what's the relationship with street and my courtyard... so, this is for me the most.

Your age: 27

Your occupation: PhD student in Urban Planning and Policy

1. From your use experience, how different do you think between typical face-to-face design consulting with an architect and the web-based home design approach like this?

   Face-to-face or internet web page, to me, they are same thing. They are the same service for the clients but the interface using different way. Architect, designer or consultant, could give clients the suggestion based on their experience, they know what clients want. However, from the exploring of the web, people know what they want and just do select. This is the difference in between face-to-face and internet, design and selection.

2. Do you think the method of using dynamic questionnaire to collect client's requirement and providing the matched design solutions online is possible and successful?

   I do not think it could cover all of the design. Maybe for a simple housing, it could work well because the functions are simple and clear, people is familiar with these and they knew what they need, so they can do select but not design.

3. Do you consider the i_Prefab Home (interface) navigation process is intuitive? What can be improved?

   Yes, I do. I like the interface design and that is very friendly and very easy for the users. But, I am not very clear what kind of clients will use this. For example, if I am not a developer but just a normal user who just want to buy a house for his/her family, I do not care how to build it and what kind of detail for construction. Perhaps the only things I will concern are the location of my house, is it convenient to parking or easy to the public transportation, or is it close to the supermarket? What kind materials, finishing? How much for those? To me, is it affordable? Maybe, add the option for the materials or typologies for windows or doors and price.

4. As a consumer, if you have a chance to participate in your home design process from the beginning, what is the most important issue from the following options and why? (1) List of the required spaces; (2) Plan layout; (3) Individual Room Design; (4) Finishes and Appliances.

   Plan layout and individual room design. Layout is for the housing function, based on that I will know how to use the room and how to use it well. The individual design will regard to individual people within the personal background, experience, that could be variety and different.

Your age: 32

Your occupation: Architect

1. From your use experience, how different do you think between typical face-to-face design consulting with an architect and the web-based home design approach like this?

    In the typical face-to-face design consulting with an architect, you can really see a person who is physically sitting in front of you and explaining everything to you. However, the architect might change your ideas (even you have your own ideas) to somewhere based on he or she prefers to do or willing to do (call architect's style). Every architect has his/her specialty, and you may or may not agree with him and waste time in arguing with the architect. Sometimes, architects will run over-budget in order to finish the "high-quality" of their design works. Opposite to that, the web-based home design provided cost estimate might be more close to the real final cost. You can play the web-based home design interface before you really want to build the home without any pressure. If you want to change something, you can always go to the website and modify the program without calling the architect in the midnight.

2. Do you think the method of using dynamic questionnaire to collect client's requirement and providing the matched design solutions online is possible and successful?

    Yes, I like this method so much. Definitely it will work one day. The only thing I am suspecting is how can you (prefab system provider/ interface designer) tell people's mind to suggest the matched design solutions? Maybe you can think about the concept of Wikipedia to allow registered users or anyone to collaboratively create, edit, link, and organize the content of a website. So, the matched design solutions will be more close to popular taste and avoid some personal bias.

3. Do you consider the i_Prefab Home (interface) navigation process is intuitive? What can be improved?

    The interface is great, clear, and user-friendly. It covers many details. As I mentioned in last question, if you can integrate with the concept of Wikipedia, the website can be extended to have some information with vendors, suppliers, and it may be good as a market place.

4. As a consumer, if you have a chance to participate in your home design process from the beginning, what is the most important issue from the following options and why? (1) List of the required spaces; (2) Plan layout; (3) Individual Room Design; (4) Finishes and Appliances.

    All above are equally important to me!

Your age: 35

Your occupation: Housewife

BIBLIOGRAPHY

Arieff, Allison and Bryan Burkhart. <u>Prefab</u>. Salt Lake City: Gibbs Smith, 2002.

Autodesk. "The i-drop® Web Publishing Framework: Publishing Intelligent Design Content on the Internet." <u>Autodesk White Paper</u> 2003.

Bahamon, Alejandro. <u>Prefab</u>. New York: HBI, 2002.

Baldwin, James T. <u>Bucky Works : Buckminster Fuller's Ideas for Today</u>. New York: John Wiley & Son, Inc., 1996.

Barrow, Larry, Shaima Al Arayedh, and Shilpi Kumar. "Performance House I: A CADCAM Modular House System." <u>Proceedings of the 2006 Annual Conference of the Association for Computer-Aided design in Architecture (ACADIA)</u>. Louisville, Kentucky, USA, 2006: 104-121.

Beaudin, Jennifer S. <u>From Personal Experience to Design: Externalizing the Homeowner's Needs Assessment Process</u>. Master Thesis, Massachusetts Institute of Technology, 2003.

Bernhardt, Arthur D. <u>Building Tomorrow: The Mobile/ Manufactured Housing Industry</u>. Cambridge: The MIT Press, 1980.

Botha, Marcel and Lawrence D. Sass. "The Instant House: Design and Digital Fabrication of Housing for Developing Environments." <u>Proceedings of the 11th International Conference on Computer Aided Architectural Design Research in Asia (CAADRIA)</u>. Kumamoto, Japan, 2006: 209-216.

Bowen, Ted S. "From Balloon Frame to CAD-tailored: Japanese Precut Framing Technology Comes to the U.S." <u>Architectural Record</u> 12 Oct. 2004 <http://archrecord.construction.com/features/digital/archives/0410dignews-1.asp>.

Brand, Stewart. <u>How Buildings Learn: What Happens After They're Built</u>. New York: Penguin, 1994.

Burkhart, Bryan, Phil Noyes, and Allison Arieff. <u>Trailer Travel: A Visual History of Mobile America</u>. Salt Lake City: Gibbs Smith, 2002.

Carper, James. <u>Using Building Systems: Modular, Panelized, Log, Dome</u>. Washington DC: National Association of Home Builders of the United States, 1990.

Cavieres, Andrés P. and Marcelo G. Quezada. "Analysis of The Possibilities Offered by the Application of Parametric Modeling Technologies in the Design Processes Shared Between Architects and Industrial Designers: The Prefabricated House Case." <u>Proceedings of the 9th Iberoamerican Congress of Digital Graphics (SIGRADI)</u>. Lima, Peru, 2005: 300-303.

Chien, Sheng-Fen and Shen-Guan Shih. "Design through Information Filtering: A Search Driven Approach for Developing a Layperson's CAAD Environment." Proceedings of the Ninth International Conference on Computer Aided Architectural Design Futures (CAAD Futures). Eindhoven, Netherlands, 2001: 103-110.

Cohen, Jonathan. "Integrated Practice and the New Architect: Keeper of Knowledge and Rules." AIArchitect Oct. 2005. 10 Mar. 2007 <http://www.aia.org/aiarchitect/ thisweek05/tw1028/tw1028bp_change_cohen.cfm>.

Cook, Peter. Archigram. New York: Princeton Architectural Press, 1999.

Combes, L. and A. Bellomio. "Creativity and Modularity in Architecture." Added Value of Computer Aided Architectural Design (AVOCAAD) 1999 Conference Proceedings. Brussels, Belgium, 1999: 169-179.

Crayton, Tim. "The Design Implications of Mass Customization." Design Intelligence Vol. 7 No 5, May 2001.

Davies, Colin. The Prefabricated Home. London: Reaktion Books, 2005.

Delary, Denise N. A Flexible and Affordable System of Prefabricated Modular Housing. Master Thesis, Illinois Institute of Technology, 2004.

Design Museum. "R. Buckminster Fuller." 1 December, 2006 <http://www.designmuseum.org/design/r-buckminster-fuller >.

Donath, Dirk and Luis F. González. "Integrated Planning Support System for Low-Income Housing." Proceedings of the 5th Iberoamerican Congress of Digital Graphics (SIGRADI). Concepcion, Chile, 2001: 113-116.

Duarte, Jose P. "A Discursive Grammar for Customizing Mass Housing - The Case of Siza´s Houses at Malagueira." Proceedings of the 21st International Conference on Education in Computer Aided Architectural Design in Europe (eCAADe). Graz, Austria, 2003: 665-674.

Duarte, Jose P. Customizing Mass Housing: A Discursive Grammar for Siza's Malagueira Houses. Ph.D. in Architecture Thesis, Massachusetts Institute of Technology, 2001.

Duerk, Donna P. Architectural Programming: Information Management for Design. New York: Van Nostrand Reinhold, 1993.

Fetters, Thomas T. The Lustron Home: The History of A Postwar Prefabricated Housing Experiment. Jefferson, N.C.: McFarland & Company, Inc., 2002.

Fortmeyer, Russell. "Loblolly House: In Stock and Ready to Ship – KieranTimberlake Associates Fabricates The Loblolly House in a Warehouse, Setting up a New Supply Chain and Establishing a Base to Revolutionize The Firm's Future Production." Architectural Record Nov. 2006: 185-192.

Friedman, Avi. The Adaptable House: Designing Homes for Change. New York: McGraw-Hill Companies, Inc., 2002.

Fröst, Peter. "A Real Time 3D Environment for Collaborative Design." Proceedings of the 10th International Conference on Computer Aided Architectural Design Futures (CAAD Futures). Tainan, Taiwan, 2003: 203-212.

Fure, Adam and Karl Daubmann. "HouseMC - Mass-Crafting Numerical Instructions for Construction." Proceedings of the 2003 Annual Conference of the Association for Computer Aided Design In Architecture (ACADIA). Indianapolis, Indiana, USA, 2003: 432-433.

Garber, Richard and Nicole Robertson. "The Pleated Cape: From the Mass-Standardization of Levittown to Mass Customization Today." Proceedings of the 2006 Annual Conference of the Association for Computer Aided Design In Architecture (ACADIA). Louisville, Kentucky, USA, 2006: 426-439.

Gausa, Manuel. Housing: New Alternatives, New Systems. Basel, Switzerland: Birkhauser, 1998.

Gershenfeld, Neil. Fab: The Coming Revolution on Your Desktop – from Personal Computers to Personal Fabrication. New York: Basic Books, 2005.

Hall, Peter. "Living for Tomorrow." Metropolis Dec. 2002. 10 October, 2005 <http://www.metropolismag.com/html/content_1202/mit>.

Hart, Sara. "Prefabrication, the Speculative Builder's Tool, Has Been Discovered by Modernist Designers: Architects Are Investigating Ways to Capture an Unserved Market for Residential Design." Architectural Record, Dec. 2003: 123-130.

Hecker, Douglas. "Dry-In House: A Mass Customized Affordable House for New Orleans." Proceedings of the 10th Iberoamerican Congress of Digital Graphics (SIGRADI). Santiago, Chile, 2006: 359-362.

Herbert, Gilbert. The Dream of the Factory-Made House. Cambridge: The MIT Press, 1984.

Hershberger, Robert G. Architectural Programming and Predesign Manager. New York: McGraw-Hill Companies, Inc., 1999.

Hinnant, Lori. "Modern Designs Find Their Way into Prefab Homes." Chicago Tribune 12 Apr. 2005.

History of Misawa – Misawa Homes Group. 10 March, 2008 <http://www.misawa.co.jp/en/info/history.html>.

Howeler, Eric. "Soft Serve: Better Living through Software." Praxis Issue 3, 2001.

Huang, Chuen-huei and Robert J. Krawczyk. "A Choice Model of Consumer Participatory Design for Modular Houses." Proceedings of the 25th International Conference on

Education in Computer Aided Architectural Design in Europe (eCAADe). Frankfurt, Germany, 2007: 679-686.

Huang, Chuen-huei. "A Choice Model of Mass Customized Modular Housing by Internet Aided Design." Joint Conference Proceedings of International Mass Customization Meeting 2007 (IMCM'07) and International Conference on Economic, Technical and Organizational Aspects of Product Configuration Systems (PETO'07). Hamburg, Germany, 2007: 61-75.

Huang, Chuen-huei. "Decision Support System for Modular Houses." Proceedings of the 12th International Conference on Computer Aided Architectural Design Research in Asia (CAADRIA). Nanjing, China, 2007: 675-676.

Huang, Chuen-huei. and Robert J. Krawczyk. "i_Prefab Home: Customizing Prefabricated Houses by Internet-Aided Design." Proceedings of the 24th International Conference on Education in Computer Aided Architectural Design in Europe (eCAADe). Volos, Greece, 2006: 690-698.

Huang, Chuen-huei, Robert J. Krawczyk, and George Schipporeit. "Integrating Mass Customization with Prefab Housing." Proceedings of the Second International Conference of Arab Society for Computer Aided Architectural Design (ASCAAD). Sharjah, UAE, 2006: 124-136.

Huang, Chuen-huei, Robert J. Krawczyk, and George Schipporeit. "Mass Customizing Prefab Modular Housing by Internet-aided Design." Proceedings of the 11th International Conference on Computer Aided Architectural Design Research in Asia (CAADRIA). Kumamoto, Japan, 2006: 203-208.

Huber, Benedikt and Jean-Claude Steinegger. Jean Prouve: Prefabrication – Structures and Elements. New York: Praeger Publishers, Inc., 1971.

Irwin, Robert. Tips and Traps When Building Your Home. New York: McGraw-Hill Companies, Inc., 2001.

Kendall, Stephen and Jonathan Teicher. Residential Open Building. New York: E & FN Spon, 2000.

Khemlani, Lachmi. "Prefabrication of Timber Buildings based on Digital Models: A Perspective from Norway." AECbytes Feb. 2005. 14 Oct. 2006 <http://www.aecbytes.com/feature/2005/Norway_prefab.html>.

Kieran, Stephen and James Timberlake. Refabricating Architecture: How Manufacturing Methodologies are Poised to Transform Building Construction. New York: McGraw-Hill Companies, Inc., 2004.

Kim, Jin B. and Yongwook Jeong. "Collaborative CAAD: State-of-the-Art and the Future – A Comparative Study of CAAD, Product Development, and Group Support Systems." Proceedings of the 9th International Conference on Computer Aided Architectural Design Research in Asia (CAADRIA). Seoul, Korea, 2004: 117-132.

Kolarevic, Branko. Architecture in the Digital Age: Design and Manufacturing. New York: Spon Press, 2003.

Kronenburg, Robert. Houses in Motion: The Genesis, History and Development of the Portable Building. Chichster, UK: Wiley-Academy, 2002.

Kronenburg, Robert. Spirit of The Machine: Technology as an Inspiration in Architectural Design. Chichster, UK: Wiley-Academy, 2001.

Kunz, Martin N. and Michelle Galindo. Best Designed Modular Houses. Ludwigsburg, Germany: Avedition GmbH, 2005.

Larson, Kent, Tyson Lawrence, Thomas J. McLeish, Deva Seetharam, and H. Shrikumar. "A Network for Customizable + Reconfigurable Housing." Proceedings of Home Oriented Informatics and Telematics (HOIT). Irvine, California, 2003. 14 Aug. 2005 <http://www.crito.uci.edu/noah/HOIT/2003papers.htm>.

Larson, Kent, Stephen Intille, Thomas J. McLeish, Jennifer Beaudin, and Reid Williams. "Open Source Building: Reinventing Places of Living." BT Technology Journal Vol. 22 No. 4, October 2004: 187-200.

Larson, Kent, Mark A. Tapia, and Jose P. Duarte. "A New Epoch: Automated Design Tools for the Mass Customization of Housing." A+U Vol. 366, March 2001: 116-121.

Lawrence, Tyson T. Chassis + Infill: A Consumer-Driven, Open Source Building Approach for Adaptable, Mass Customized Housing. Master Thesis, Massachusetts Institute of Technology, 2003.

Leupen Bernard. Frame and Generic Space. Rotterdam: 010 Publishers, 2006.

Live in your next Toyota – Buying a House – MSN Real Estate. 10 March, 2008 < http://realestate.msn.com/buying/articlenewhome.aspx?cp-documentid=596034>.

NAHB Research Center, Inc. Factory and Site-Built Housing: A Comparison for the 21st Century. Washington DC: U.S. Department of Housing and Urban Development, 1998.

NAHB Research Center, Inc. Model Guidelines for Design, Fabrication, and Installation of Engineered Panelized Walls. Washington DC: U.S. Department of Housing and Urban Development, 2001.

Negroponte, N. Soft Architecture Machines. Cambridge: The MIT Press, 1975.

Noguchi, Masa. "Mass Custom Home: The Mass Custom Design Approach to the Delivery of Quality Affordable Homes." 13 Sep. 2003 <http://www.masscustomhome.com>.

Nylander, Ola. Architecture of The Home. Chichster, UK: Wiley-Academy, 2002.

Peña, William M. and Steven A. Parshall. Problem Seeking: An Architectural Programming Primer. New York: John Wiley & Sons, Inc., 2001.

Pine, B. Joseph II. Mass Customization: The New Frontier in Business Competition. Boston: Harvard Business School Press, 1993.

Ragon, Michel. Goldberg: Dans La Ville/ On The City. Paris: Paris Art Center, 1985.

Rahman, Rashidah and Alan Day. "A Comparative Study of Digital and Traditional Tools for Participative Design." Proceedings of the Second International Conference of Arab Society for Computer Aided Architectural Design (ASCAAD). Sharjah, UAE, 2006: 233-252.

Rizal, H. and Ahmad Rafi. "The Impact of Internet Enabled Computer Aided Design (iCAD) in Construction Industry." Proceedings of the 7th International Conference on Computer Aided Architectural Design Research in Asia (CAADRIA). Cyberjaya, Malaysia, 2002: 85-92.

Rodker, John. Towards a New Architecture by Le Corbusier. New York: Dover Publications, Inc., 1986. (Reprint) Originally published: London: J. Rodker, 1931

Sanoff, Henry. Methods of Architectural Programming. Stroudsburg, Pennsylvania: Dowden, Hutchinson & Ross, Inc., 1977.

Sarid, A. and R. Oxman. "The Web as a Knowledge Representational Media for Architectural Precedents." Proceedings of the 18th International Conference on Education in Computer Aided Architectural Design in Europe (eCAADe). Weimar, Germany, 2000: 245-249.

Sherwood, Roger. Modern Housing Prototypes. Cambridge: Harvard University Press, 1978.

Schmitt, Gerhard. Information Architecture: Basis and Future of CAAD (The Information Technology Revolution in Architecture). Basel, Switzerland: Birkhauser, 1999.

Schodek, Daniel, Martin Bechthold, Kimo Griggs, Kenneth M. Kao, and Marco Steinberg. Digital Design and Manufacturing: CAD/CAM Applications in Architecture and Design. Hoboken, New Jersey: John Wiley & Sons, Inc., 2004.

Siegal, Jennifer. Mobile: The Art of Portable Architecture. New York: Princeton Architectural Press, 2002.

Steven Winter Associates, Inc. Next Generation of Manufactured Housing Design Phase. Washington DC: U.S. Department of Housing and Urban Development, 1997.

Stouffs, Rudi, Bige Tunçer, and Sevil Sariyildiz. "The Customer is King: Web-Based Custom Design in Residential Developments." Proceedings of the Sixth Conference on Computer Aided Architectural Design Research in Asia (CAADRIA). Sydney, Australia, 2001: 149-157.

Stover, Dawn. "Building Blocks." Popular Science Magazine Nov. 2006: 78-84.

Taliesin Associated Architects. Production Dwelling: An Opportunity for Excellence. Spring Green, Wisconsin: Taliesin Associated Architects of the Frank Lloyd Wright Foundation, 1970.

Thillart, C. C. A. M. Customised Industrialisation in the Residential Sector: Mass Customisation Modelling as a Tool for Benchmarking, Variation, and Selection. Amsterdam: SUN Publishers, 2004.

Thornton, Rosemary. The Houses That Sears Built. Alton, IL: Gentle Beam Publications, 2004.

TOYOTA: Non-Automotive – Housing. 10 March, 2008 <http://www.toyota.co.jp/en/more_than_cars/housing/index.html>.

Ulrich, Karl and K. Tung. "Fundamentals of Product Modularity." Proceedings of the 1991 ASME Winter Annual Meeting Symposium on Issue in Design/Manufacturing Integration, Atlanta, 1991: 73-79.

Urban Designers Machiya (プ ラ ン) | ミ サ ワ ホ ー ム . 1 April, 2008 <http://www.misawa.co.jp/kodate/syouhin/mokusitu/urban_designers_matiya/special/plan_outvoid.html>.

Wachsmann, Konrad. The Turning Point of Building. New York: Reinhold Publishing Corporation, 1961.

Wachsmann, Konrad. Building the Wooden House. Basel, Switzerland: Birkhauser, 1995.

Wallis, Allan D. Wheel Estate: The Rise and Decline of Mobile Homes. New York: Oxford University Press, 1991.

Warszawski, Abra. Industrialized and Automated Building Systems: A Managerial Approach. London: E & FN Spon Press, 1999.

Wentling, James. Housing by Lifestyle: The Component Method of Residential Design. New York: McGraw-Hill Companies, Inc., 1995.

Zhou, Qi "From Cad to iAD: A Prototype Simulation of the Internet-based Steel Construction Consulting for Architects." Proceedings of the 8th International Conference on Computer Aided Architectural Design Research in Asia (CAADRIA). Bangkok, Thailand, 2003: 919-936.

Zhou, Qi, Robert J. Krawczyk, and George Schipporeit. "From Cad to iAD: A Web-based Steel Consulting of Steel Construction in Architecture." Proceedings of the 20th International Conference on Education in Computer Aided Architectural Design in Europe (eCAADe). Warsaw, Poland, 2002: 346-349.

Zhou, Qi, Robert J. Krawczyk, and George Schipporeit. "From Cad to iAD: A Working Model of the Internet-Based Engineering Consulting in Architecture." Proceedings of the 7th International Conference on Computer Aided Architectural Design Research in Asia (CAADRIA). Cyberjaya, Malaysia, 2002: 73-80.